# Evaluating the Reliability of Emergency Response Systems for Large-Scale Incident Operations

Brian A. Jackson, Kay Sullivan Faith, Henry H. Willis

Prepared for the Federal Emergency Management Agency

RAND
HOMELAND SECURITY AND DEFENSE CENTER

This research was sponsored by the Federal Emergency Management Agency and was conducted under the auspices of the RAND Homeland Security and Defense Center, a joint center of the RAND National Security Research Division and RAND Infrastructure, Safety, and Environment.

**Library of Congress Cataloging-in-Publication Data**

Jackson, Brian A., 1972-
Evaluating the reliability of emergency response systems for large-scale incident operations / Brian A. Jackson, Kay Sullivan Faith, Henry H. Willis.
p. cm.
Includes bibliographical references.
ISBN 978-0-8330-5005-2 (pbk. : alk. paper)
1. Emergency management—United States—Evaluation. 2. Preparedness—Evaluation. 3. Incident command systems—United States. 4. Assistance in emergencies—United States. I. Faith, Kay Sullivan. II. Willis, Henry H. III. Title.

HV551.3.J328 2010
363.34'80684—dc22

2010024680



Published 2010 by the RAND Corporation
1776 Main Street, P.O. Box 2138, Santa Monica, CA 90407-2138
1200 South Hayes Street, Arlington, VA 22202-5050
4570 Fifth Avenue, Suite 600, Pittsburgh, PA 15213-2665
RAND URL: http://www.rand.org/
To order RAND documents or to obtain additional information, contact
Distribution Services: Telephone: (310) 451-7002;
Fax: (310) 451-6915; Email: order@rand.org

# Preface

The ability to measure emergency preparedness—to predict the likely performance of emergency response systems at future events—is critical for policy analysis in homeland security. It is also key for answering the fundamental question that the public and policymakers alike have about those systems: How much confidence should we have that they will function as planned when the next large-scale incident or disaster occurs? Though substantial effort has been devoted to developing measures of preparedness in a range of fields, good measures are still elusive. This work makes a contribution to that larger effort, by drawing on the fields of systems analysis and engineering and applying concepts of system reliability to the evaluation of response systems. By laying out a planned response operation in detail and systematically asking what might go wrong that will *prevent* the response system from performing as designed, this approach can help to estimate the likelihood that the response system will be able to meet the needs of a future large-scale incident or disaster.

This work was sponsored by the Federal Emergency Management Agency, National Preparedness Directorate, National Preparedness Assessment Division, as part of a larger project carried out by the Center for Risk and Economic Analysis of Terrorism Events (CREATE), a Department of Homeland Security Center of Excellence based at the University of Southern California.

This work should be of interest to individuals at the federal, state, and local level involved in preparedness and planning; members of the private sector involved in contingency and business continuity planning; members of the executive and legislative branches interested in homeland security, emergency management, assessment, and performance measurement; and members of the public interested in disaster and emergency preparedness. Related RAND publications include the following:

- Brian A. Jackson, *The Problem of Measuring Emergency Preparedness: The Need for Assessing "Response Reliability" as Part of Homeland Security Planning*, 2008.
- Henry H. Willis et al., *Initial Evaluation of the Cities Readiness Initiative*, 2009.
- Christopher Nelson et al., *New Tools for Assessing State and Local Capabilities for Countermeasure Delivery*, 2009.
- Tom LaTourrette et al., *Public Health Preparedness and Response to Chemical and Radiological Incidents: Functions, Practices, and Areas for Future Work*, 2009.

## The RAND Homeland Security and Defense Center

This research was conducted under the auspices of the RAND Homeland Security and Defense Center, which conducts analysis to prepare and protect communities and critical infrastructure from natural disasters and terrorism. Center projects examine a wide range of risk management problems, including coastal and border security, emergency preparedness and response, defense support to civil authorities, transportation security, domestic intelligence programs, technology acquisition, and related topics. Center clients include the Department of Homeland Security, the Department of Defense, the Department of Justice, and other organizations charged with security and disaster preparedness, response, and recovery. The Homeland Security and Defense Center is a joint center of the RAND National Security Research Division and RAND Infrastructure, Safety, and Environment.

Questions or comments about this monograph should be sent to the principal author, Brian A. Jackson (Brian_Jackson@rand.org). Information about the Homeland Security and Defense Center is available online (http://www.rand.org/multi/homeland-security-and-defense/). Inquiries about homeland security research projects should be sent to:

Andrew Morral, Director
Homeland Security and Defense Center
RAND Corporation
1200 South Hayes Street
Arlington, VA 22202-5050
703-413-1100, x5119
Andrew_Morral@rand.org

# Contents

# Figures

# Tables

# Summary

Societies build emergency response systems to be there when damaging incidents—whether natural or caused by man—occur. Though the effectiveness of those systems in responding to everyday emergencies is easy to see, knowing how prepared they are to deal with large-scale incidents—which, fortunately, are far rarer—is much more difficult. Most of the time, responses to even large-scale emergencies go very well. But sometimes they do not, leading to questions about why response activities did not go as expected and what policy actions should be taken in response.

Practitioners and researchers in many fields have devoted significant effort to developing ways to measure emergency preparedness. Progress has been made—in the creation of systems to assemble data on preparedness inputs, national policy documents that begin to set standards for capability levels, and exercises designed to test preparedness systems—but the ability to measure preparedness has still been highlighted as an area requiring attention and innovation (FEMA, 2009b). This work helps address that shortfall by approaching preparedness assessment from a perspective that is very different from those used in most previous efforts.

We view the response operation for a large-scale emergency or disaster as a system, one that is built to address post-incident needs and potentially involves multiple separate organizations.[1] In this view, a response system is defined by a set of plans, resources, authorities, agencies, and their associated human resources. We draw on tools from the systems analysis and engineering fields for analyzing system performance as a way of looking at potential response performance at future incidents. This includes laying out what the system is intended to do and exploring what inputs and resources are required for it to deliver, a component common to most preparedness assessment efforts. But we also look at the system and assess what might go wrong—what breakdowns or "failure modes" might threaten the ability of the system to perform effectively. This second part

[1] This framing is consistent with the Emergency Management Accreditation Program's definition of *emergency management program* as a "jurisdiction-wide system that provides for management and coordination of prevention, mitigation, preparedness, response and recovery activities for all hazards" (EMAP, 2007). Such a system "encompasses all organizations, agencies, departments, entities and individuals responsible for emergency management and homeland security functions," though the focus in our work was on the system's preparedness and response activities.

is the heart of our analysis and distinguishes our approach from most preparedness assessment methods. The combination of these two approaches can help to answer the fundamental question that the public and policymakers have about response systems: *How much confidence should we have that they will function as planned when the next large-scale incident or disaster occurs?*

## Assessing the Reliability of Response Systems

To answer that question, what is needed is a measure of the likelihood that a response system will perform well—or, put another way, that events that prevent it from performing well will not occur—at a future incident. We have labeled that measure *response reliability*, the probability that a response system will be able to deliver at or above a given level of capability at a future emergency incident. Our framing of response reliability takes into account that the scale, scope, and complexity of an incident matters. A given response system may perform predictably well for events up to a certain level of required performance, but above that level problems may considerably reduce the chances that the system will be able to meet the requirements of the incident.

To evaluate response reliability, we adapted analytical techniques from the reliability analysis and risk assessment fields—specifically, fault tree analysis and failure mode, effects, and criticality analysis (FMECA). Building on the ways these techniques are applied to technical systems, we framed four steps for analysis of response systems for large-scale incidents (and diagram them in Figure S.1).

**Figure S.1**
**The Four Steps of Response Reliability Analysis**

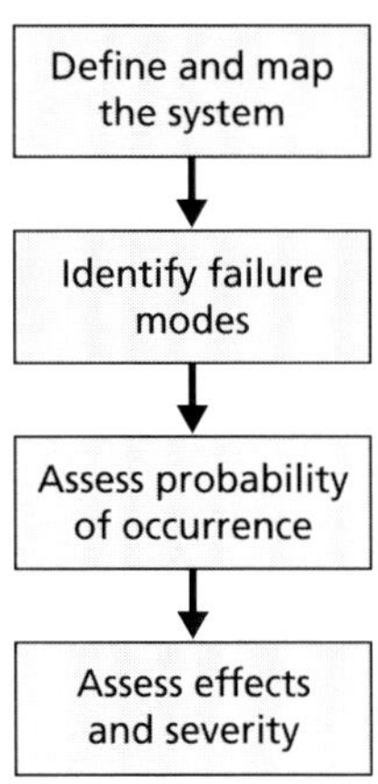


RAND *MG994-S.1*

1. **Define and Map the System.** Understanding what might go wrong in a system requires knowing how it is put together. Laying out the different functions (in the case of response operations) that must be performed and how they link together defines the structure and bounds of the analysis. For example, evacuating an area requires transporting people who have no transportation of their own, which involves not just vehicles and drivers but also (1) responders to manage gathering the people and their orderly boarding, (2) public communications capability to get the message out to people that an evacuation is under way, (3) information collection capabilities to identify where the people who need assistance are, and (4) a functioning incident management system to fit all the activities together and make sure that they are completed in time to be valuable. Breakdowns in the system could be caused either within individual components or at the seams between components that depend on one another. Defining the system requires understanding what it means for each part of the system to work well and determining how reductions in performance would affect outcomes.
2. **Identify Failure Modes.** Failure modes are defined as "the observable manners in which a component fails" (Ebeling, 1997, p. 168), which in this case would be the ways that performance of different parts of the response system would break down. Examples of failure modes for response activities include staffing problems, human errors, equipment breakdowns (e.g., damage to response vehicles), and so on. Some failures might occur as a response operation was being initiated, while others might occur at later points in the response. Failures may be due to random events (e.g., equipment failures even though all appropriate maintenance had been done), have a clear human cause (e.g., maintenance had not been done), or be caused by external events (e.g., the incident damaged the vehicles prior to deployment). In our work, we drew on the practitioner literature, response after-action reports (AARs), past RAND research, and other published sources to help identify the ways that response operations can break down. We account for the effect of failure modes on each part of the system by determining whether each mode is specific to one response function or capability or has more general effects on multiple parts of the system.
3. **Assess the Probability of Occurrence of Different Failure Modes.** Given many things that could hurt the functioning of a system, one differentiator among them is how likely they are to happen. The probability that a specific failure will occur during a response operation could be estimated a number of different ways; for example, the estimate might be based on real-world data on the occurrence of failures in past responses, or it might be based on estimates elicited from practitioners or subject-matter experts. Different ways of estimating the probability of failure have their own strengths and weaknesses with respect to accuracy and ease of implementation. Depending on how failure

modes have been defined, some calculation may be involved in determining the probability of a specific mode. For example, if the failure mode of concern for response is a communications system breakdown and there are both primary and backup systems, then the probability of the failure would be the probability *both* systems failed.

4. **Assess the Failure Mode Effects and Their Severity.** Other differentiators among failure modes are their effect and severity. In FMECA, this assessment is done at the system level, by asking, "What is the effect of the failure mode's occurrence on overall system performance?" Different events can have a variety of effects. In considering the effect of failures on emergency response operations, we viewed failures as falling into two main classes: (1) response–termination failures—that is, failures that would stop a response operation entirely—and (2) capability–reduction failures—that is, failures that make a response operation less effective but do not halt response (e.g., an event that reduces the number of victims a hospital can accept after an incident). Effectiveness-reduction failures may cause a reduction in system performance either directly or via their effects on other response functions—for example, difficulties implementing incident command could have an effect on many other response activities. The severity of an effectiveness-reduction failure is determined by the size of its impact relative to the requirements of the response operation.

FMECA is one of a number of methods in reliability analysis for combining information on the incidence and consequence of failures into an overall assessment of a system. For our work, we found this process attractive because the approximations that are made allow each failure mode to be treated independently, somewhat simplifying use of the technique for assessing a complex response system.

In our study, we explored FMECA from two complementary perspectives.

First, we did a complete theoretical analysis of a very simple response system to fully demonstrate the methods and show how both qualitative and quantitative assessment of response reliability could help inform preparedness planning and resource allocation. Statistical modeling of a response operation affected by a variety of different failure modes made it possible to show how the probability that the system would perform effectively at incidents of increasing levels of required performance could be calculated. Figure S.2 illustrates how graphs of a response system's reliability (on the vertical axis) delivering increasing levels of capability or performance (moving right on the horizontal axis) can provide an overall snapshot of its likely future performance. The curves in the figure show three exemplary cases: a system that is perfectly reliable (the gray dotted line) and so functions perfectly up to the highest level of performance it has been designed for (the maximum response capacity, $RC_{max}$); a system with serious problems whose probability of good performance drops very rapidly as the required performance level increases (the red line, labeled a "brittle system"); and a

**Figure S.2**
**Illustrative Reliability Curves for Response Systems of Varied Performance**

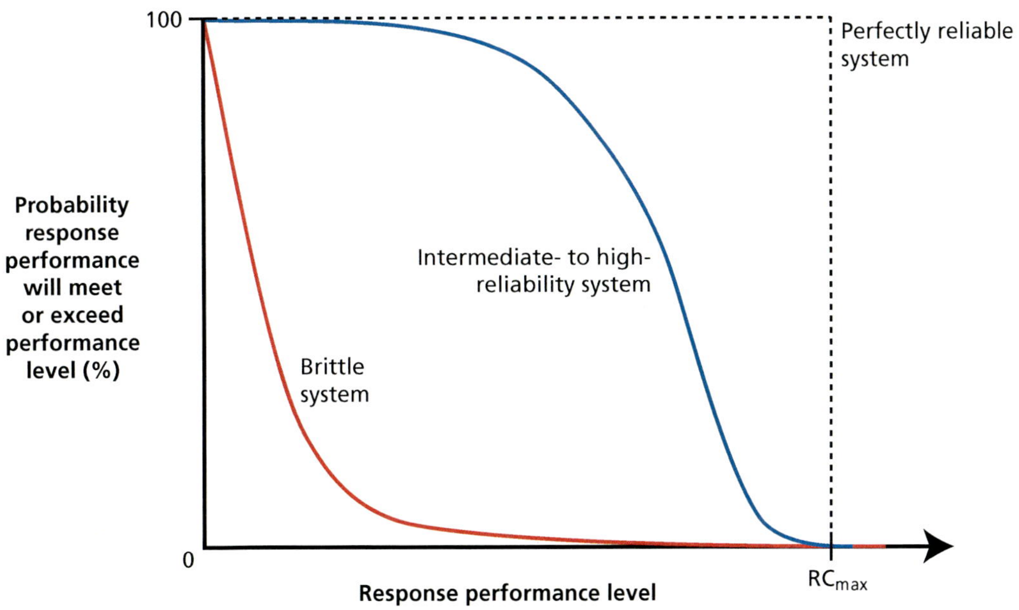


RAND *MG994–S.2*

more realistic system (the blue line) that performs well over a large range of incidents but has difficulty for those close to its maximum designed capacity. For a real response system, such an analysis could be done with sufficiently detailed results from the process described above, identifying failure modes and estimating their probabilities of occurrence and the severity of their consequences.

Second, we analyzed a more realistic response for a chlorine release incident, drawing on AARs from past response operations as a data source. This element of the work was designed as a proof-of-concept demonstration of the analysis using real-world data that response organizations already produce. We constructed a model of the response to a chlorine incident, covering the elements of all response tasks from incident management through assisting exposed victims. We identified failure modes for each part of the model, including critical interdependencies among different parts of the response operation. We then analyzed a convenience sample of 70 AARs, describing 65 separate response operations. All but two of the AARs described actual incidents, with the remainder describing exercises. We examined each AAR for the presence of different failure modes during the response operation and any descriptive information on the consequences of each failure's occurrence. This second phase of the work simultaneously demonstrated the promise and the challenge of the analytic approach when applied to real-world response systems.

## Using the Results of Reliability Assessment in Preparedness Evaluation and Decisionmaking

The goal of this work was to show that adapting techniques from reliability engineering and risk analysis for evaluating the performance of technical systems can contribute to better ways of evaluating preparedness and anticipating the likely future performance of emergency response systems in large-scale events. We believe that we have achieved that goal, and have demonstrated that both the *process* of such analyses and their *results* can potentially contribute to preparedness planning and evaluation in different but complementary ways.

The first step of the process, defining and mapping the response, takes an explicitly systems-oriented approach to how an entire response operation functions. In our model, we do not distinguish which responders will perform the tasks in each part of the overall system, in terms of which organizations they are a part of or which response disciplines they are trained in. By ignoring the insignia on the uniforms of individual participants in the response, this approach lays out in black and white the potential interdependencies among organizations and how seams between them could result in response failure. In discussing our work with one response practitioner, the comment was made that "though we are supposed to be breaking stovepipes, we still do a lot of our planning within single agencies—and this captures the problems that can still create."

The second step, systematically identifying failure modes for each part of the response model, provides a structured method for doing the type of "what-if" questioning done by experienced planners, and also for capturing the results of that process so they can be explicitly included in an organization's plan and the knowledge spread among its staff. Working down to the level of individual root failure modes also makes it easier to identify solutions to problems that are discovered, since different failure modes—even ones within the same response function—can have very different "fixes." Even just counting up failure modes and determining the breadth of their effects can help inform prioritization, with failure modes that have broad effects on performance standing out as causes for particular concern.

The third and fourth steps—assessing the probability, effects, and severity of the consequences of individual failure modes—get at the information needed to identify specific priorities and to assess the value of different preparedness investments. In our work, we drew on existing AARs from response operations to test this part of the analysis with real-world data. The AARs we examined proved to be a challenging data source. But we were nevertheless able to apply the basic analytical process we describe, and this process made it possible to extract useful data from a very heterogeneous dataset. Though we were seeking that data to inform qualitative and quantitative measures for response performance, practitioners who we interacted with also suggested other uses for such datasets. For example, for a specific jurisdiction, data showing that failures were adding up in a specific area could be used as a way to suggest which parts of the response system might need "preventive maintenance"—refreshers in training, particular focus in near-term exercises, and so on—to reduce their chances of recurrence in the future. Such applications could help to address requirements for exercises and corrective action programs in relevant emergency management standards (e.g., EMAP, 2007; NFPA, 2007).

In considering potential future implementation of these methods for broader preparedness assessment, a variety of other data sources may be superior to AARs for providing the information needed. Some such systems—for example, current preparedness assessment systems and remedial action management programs at the national level (FEMA, 2009b, p. ii) or local equivalents—might provide even better data on specific failure modes and their consequences, which could inform higher-resolution analysis of real response systems. These methods have the potential to contribute to current efforts to improve preparedness assessments (such as those required by the Post-Katrina Emergency Management Reform Act [P.L. 109-295]). Similarly, though our proof-of-concept work here used historical data from AARs, these approaches could be applied to more real-time datasets on response performance. Doing so would be consistent with the Federal Emergency Management Agency's goal to "support a living reporting mechanism that will provide an up-to-date resource on the current state of preparedness" (FEMA, 2009b, p. 1) in the nation.

Comparing the results of our reliability analysis of a real-world response operation (using AARs) with our illustrative analysis of a simple response system using simulated data, we could not take our analysis as far "in practice" as we could "in theory." In part, this was due to shortcomings in the AARs as a data source; small changes in the type of information included in such documents—i.e., capturing some estimate of the seriousness of the consequences of response issues in AARs—could make them much more useful for this type of analysis. Nonetheless, the results of our analysis and simulation using a simpler response scenario demonstrate the broader potential of reliability analysis to contribute to preparedness planning and evaluation. Though the data available to us did not support highly quantitative analysis of the chlorine response scenario, to the extent that response reliability curves can actually be estimated for real-world response operations, they could help provide a direct answer to the question—"What is the chance that things will work next time?"—that most current preparedness assessment methods cannot.

Having such a measure would help to inform policy debate of preparedness issues in a number of ways. Quantifying response reliability would help to make clear how much reliability the public should expect given current investments in preparedness, clearly articulate the cost of increasing it, and provide a means to compare different possible investments to do so—from surgically fixing known failure modes to just buying more capability to put additional slack into the system to respond to an unknown future. Reliable data on or solid estimates of response systems' reliability would help to focus preparedness policy debate and inform consideration of the truly key questions in this area: not just "How much money should we spend?" but "How exactly should we spend it?" and not just "Do we need to spend more?" but "How do we know when we have invested enough?"

## Acknowledgments

Like many projects, the success of this effort depended on the efforts of a number of individuals beyond the research team. Two of our RAND colleagues, Jeremiah Goulka and Angel Martinez, made key contributions at various points in the work. At CREATE, we would like to gratefully acknowledge the early contributions of Tony Barrett, whose past work on chlorine incidents and responses informed how we carried out our study.

In developing the simulation models used in the project for exploring different scenarios related to response reliability, Anduin E. Touw made a major contribution to the effort both in providing statistical consultation and in the design and implementation of the computational models used.

We gratefully acknowledge funding from the Federal Emergency Management Agency, National Preparedness Directorate, National Preparedness Assessment Division. We are grateful for the contributions of both Sharon Kushnir and Laureen Daly, both of whom oversaw the study at various points. In her role overseeing the project's later phases, Laureen helped us to connect with key practitioners whose input to the study was very valuable. Multiple individuals at CREATE were also key in making this project possible. We particularly thank Detlof von Winterfeldt, former director of CREATE, and Isaac Maya for their support of the study from its initiation through its completion.

We also gratefully acknowledge our peer reviewers, John Halliday of RAND and William L. Waugh, Jr., of Georgia State University, for their useful input to the document. All shortcomings obviously remain the sole responsibility of the authors.

# Abbreviations

| | |
|---|---|
| AAR | after-action report |
| CREATE | Center for Risk and Economic Analysis of Terrorism Events |
| CSB | Chemical Safety and Hazard Investigation Board |
| DHS | U.S. Department of Homeland Security |
| EMS | emergency medical services |
| EOC | emergency operations center (system-level incident command) |
| EPA | U.S. Environmental Protection Agency |
| ESF | emergency support function |
| FEMA | Federal Emergency Management Agency |
| FMEA | failure mode and effects analysis |
| FMECA | failure mode, effects, and criticality analysis |
| hazmat | hazardous materials |
| HVAC | heating, ventilating and air conditioning |
| IAP | incident action plan |
| IC | incident command |
| ICR | initiation capability-reduction failure |
| IDLH | immediately dangerous to life or health |
| IRT | initiation response-termination failure |
| IMS | incident management system |
| LLIS | Lessons Learned Information Sharing System |
| NFPA | National Fire Protection Association |
| NIMS | National Incident Management System |

| | |
|---|---|
| NTSB | National Transportation Safety Board |
| PIO | public information officer |
| PPE | personal protective equipment |
| ppm | parts per million |
| $RC_{max}$ | maximum response capacity |
| $RC_{max}$failed | maximum response capacity after a failure |
| RCR | random capability-reduction failure |
| RRT | random response-termination failure |
| TCL | Target Capabilities List |

CHAPTER ONE

# Introduction: Measurement and Emergency Preparedness

Bad things happen. Natural events, such as hurricanes, wildfires, floods, and earthquakes, kill, injure, and create destruction over significant areas. Human-caused incidents, ranging from industrial accidents to deliberate acts of terrorist or criminal violence, can similarly injure or kill people, damage property, and disrupt daily life.

Recognizing that disasters will occur, we make investments in emergency preparedness. We train firefighters to deal with everything from everyday kitchen fires to wildland firefighting operations that may involve hundreds or even thousands of responders. We store relief supplies in warehouses, for delivery to flood victims who have lost their homes and are temporarily unable to care for themselves. We develop national policies and frameworks, such as the Department of Homeland Security's (DHS's) *National Incident Management System* (NIMS) and *Target Capabilities List* (TCL) to guide planning and help to integrate disparate preparedness efforts.[1] Organizations such as the National Fire Protection Association and the Emergency Management Accreditation Program develop standards to help distinguish strong from weak preparedness programs. As a society, we take myriad other steps and make substantial investments—from the community to the national level—to prepare for varied types of emergencies.

Most of the time, when disaster strikes, response systems work exactly as planned—and perform as expected. To look at one descriptive statistic, across the United States there were 59 presidentially declared disasters and 49 fire management assistance declarations for major wildfires in 2009 (FEMA, 2009a). For most readers, many (or even most) of those events likely passed without notice, since the response organizations and systems charged with responding to these emergencies did so effectively, meeting the needs of the affected individuals and areas (see discussion in Miskel, 2008).

[1] The NIMS (DHS, 2008b) was developed to provide a common management framework to organize response operations across the country. Based on similar systems developed over many decades in the wildland fire community, the NIMS structure standardizes how different functions in a response system are defined to strengthen the ability of separate response organizations to combine their efforts at an incident. The TCL (DHS, 2007b) is a national-level document that defines the capabilities involved in response (and other) operations, lays out their interconnections and dependencies, and provides some planning guidance for the levels of capabilities required for different areas.

But sometimes there are incidents for which the system does not perform as expected. To be effective, capabilities and resources have to be delivered *where* they are needed, *when* they are needed, to the *people that need them*, and these capabilities and resources have to be able to *do what is necessary* when they get there. What can prevent that from happening? Put simply, things go wrong that were not foreseen—or if they were foreseen, were not addressed—when the response agencies were putting together their plans for future responses. Perhaps a plan for getting aid into an area depended on a bridge or airport that was destroyed in the disaster. Or information on where food aid was needed couldn't be transmitted from the affected area because communications links were severed. Or agency disputes about who was in charge of response operations led to poor coordination and inefficient use of manpower. Things went wrong and the response didn't go as planned.

That things can go wrong during emergency and disaster response operations is not surprising. But how far they go wrong is significant. What is important is whether response resources and capabilities get where they are needed. In some cases, response organizations can adapt to problems "on the fly," with minimal effect on performance. In others, operations are derailed and available resources never make their way to those who need them. It is this second type of problem that is of greatest concern to both the public and their elected officials, who then reasonably question *why* the system did not work as expected and what (if anything) should be changed before the next disaster strikes.

## Public Expectations and Our (Imperfect) Ability to Measure Emergency Preparedness

The immediacy and tangibility of unmet needs after disasters make it easy to identify situations where more could have been done. In such instances, however, it is not always the case that response organizations could reasonably have met those needs.[2] Given a finite amount of resources put into preparedness and planning, there are limits to how much can be done, and how quickly, after a disaster. While the desire to help as many people as possible after such an incident clearly reflects the best of human intentions, it does not make the expectation of such performance realistic.

But there are also clearly situations where public outrage about inadequate performance of emergency response systems is both understandable and well placed. If situations that should reasonably have been considered in planning were overlooked, if monies devoted to preparedness were spent unwisely, or if management of a response operation was badly carried out, it is hardly surprising that the public might demand

---

[2] See discussion in Miskel, 2008, on both expressed public expectations for response and the capability of response organizations to meet them.

that the problems be fixed and that the individuals or organizations responsible for the problems be punished as well.

To distinguish these different causes and shape appropriate—and effective—responses to improve preparedness, we have to be able to *measure* emergency preparedness—to project the likely future performance of single or multiple response organizations at possible future events, given the resources provided to them and the plans that they have made. Such measurement would make it possible to demonstrate the level of performance that response systems can provide, given the efforts that a city, state, or the country as a whole have made to build emergency response capacity. Such measurement could also help to distinguish poor planning from other possible reasons for poor performance, contribute to accountability for past planning and preparedness efforts (by helping to weigh "how much" preparedness they yielded versus what was promised), and help to evaluate whether a shortfall was caused by poor implementation of an otherwise sound preparedness strategy.

How is preparedness measured now? A variety of efforts to assess preparedness have been made over the years, but most of them focus on measuring what we would broadly label the *capacity* of preparedness organizations—counting how many responders are available and how much equipment is on hand, asking whether planning activities or exercises have been held, and so on—and comparing that capacity to the assumed requirements of particular incidents or scenarios of concern.[3] These approaches are effective in evaluating the *inputs* needed for response operations[4]: whether capabilities are included in a response plan, whether there is sufficient staffing to execute particular tasks, whether training has occurred, and so on.[5] While it is certainly the case that a response could fail to achieve its desired outcome because of insufficient response capacity in the affected area, failures could also arise because the response organization(s) involved cannot successfully deliver and utilize the capacity that is available. Some other approaches seek to examine performance in actually utilizing capabilities to produce response *outcomes*, particularly preparedness exercises or simulations that are designed for the purposes of evaluation.[6] Such approaches have

---

[3] See discussion and references cited in Jackson, 2008, pp. 5–10; Nelson et al., 2007a; Nelson et al., 2007b; Willis et al., 2009. FEMA, 2009b, pp. 111–113, includes a list of current preparedness-assessment-related systems and some description of their content.

[4] The Federal Emergency Management Agency's (FEMA's) 2009 *Federal Preparedness Report* highlighted this problem, specifically stating that "measurement of progress in preparedness is often limited to assessments of the amount of resources invested towards particular goals. When comprehensive numeric outcome data are not available, this report emphasizes narrative details of preparedness" (FEMA, 2009b, p. 3).

[5] For example, in the TCL, target capability levels are generally defined in numbers of response units per area (where area can be a large city, state, region, etc.), per population, or by another measure defining a scaled *resource level* (DHS, 2007b).

[6] In contrast to, for example, exercises designed predominantly to train the participating responders.

the potential to more fully characterize response systems' likely future performance at incidents similar to the simulated conditions.

## Response Reliability as a Different Approach to Preparedness Assessment

What most current approaches to preparedness measurement appear to be missing is a way to answer the fundamental question about preparedness systems: How confident should we be that the response system will perform as expected when the next large-scale incident or disaster occurs? Framed this way, the issue is not just whether enough equipment has been bought or whether responders have been trained to operate in post-incident environments—it is whether the system as a whole *will actually work* when called on.

By putting response plans in place, hiring responders, buying supplies, training, and performing other preparedness activities, a response organization—or set of organizations within a jurisdiction, state, region, or the country as a whole[7]—builds a reservoir of capacity that is available should an incident occur. For example, when a fire or terrorist attack with many victims occurs and a jurisdiction's response plan for a mass casualty incident is activated, that response plan defines the architecture of a system of organizations, people, resources, and so on that is designed to surge resources to the scene of the incident, transport victims to medical facilities, and provide extra capacity at those facilities.

Based on the resources that have been put in place and the actions defined in the response plan, a response system[8] will have some theoretical maximum capacity to care for people injured in a mass casualty incident.[9] For example, if a mass casualty plan includes a surge in hospital beds and staff and emergency supplies to temporarily

[7] This framing is consistent with the Emergency Management Accreditation Program's definition of *emergency management program* as a "jurisdiction-wide system that provides for management and coordination of prevention, mitigation, preparedness, response, and recovery activities for all hazards" (EMAP, 2007). Such a system "encompasses all organizations, agencies, departments, entities and individuals responsible for emergency management and homeland security functions," though the focus in our work was on the system's preparedness and response activities.

[8] When we use the term *response system*, we mean the system defined by whatever response plan has been activated to address the incident that has occurred. As a result, different incidents in the same jurisdiction would likely have very different response systems associated with them. For a smaller incident, a single organization might respond, and its maximum capacity would be defined by the resources of that one agency and the plan for how they are to be used. For a larger incident, the plan might involve unified command of multiple local organizations plus local or regional mutual aid. In that case, the nature of the system would be very different, and its maximum response capacity significantly larger.

[9] As discussed above, the TCL (DHS, 2007b) provides some examples of such defined capability levels for different types of jurisdictions, areas, states, regions, and the nation overall.

double the number of patients that can be served, then the response system's maximum capacity will be twice whatever level of service those facilities can provide on any other day. If everything goes according to plan, the system should theoretically be able to deliver that maximum response capacity in every incident that occurs.

But in unexpected circumstances, things will seldom, if ever, go *exactly* according to plan. Looking at after-action reports of past response operations provides a wide range of examples of things that can go wrong, affecting response performance in different ways and to varying degrees. When the response plan in our notional mass casualty incident is activated, any number of things might go wrong. For instance, communications problems might end up resulting in patients being sent to treatment locations that are already full and therefore unable to help them. Or, disruptions in setting up incident command might delay the response, meaning that less time is available to help people.

The possibility that such events will disrupt response operations reduces the chance that the response system will be able to perform as well as planned. It follows that the potential for such problems to arise should therefore reduce the confidence that policymakers and the public have in the system's ability to deal with future emergencies. But *how much* should the potential for things to go wrong reduce confidence in the response system's future performance? Answering that question requires asking three more specific questions:

- **How likely is it that individual problems will occur?** If a response system has many things that are very likely to go wrong—e.g., its communications system and its fire trucks are so old that they break down almost every time they are used—then confidence in its future performance should be modest at best.
- **What type of effect will particular problems have on the functioning of the response system?** Some problems that can occur have relatively minor effects on response performance, while others greatly reduce—or even prevent—the response system's ability to meet needs. A system afflicted by many potential problems that have a large effect on its performance should inspire less confidence than one with few or none.
- **How do the problems that could occur affect how the system can respond to incidents of different sizes, scales, or complexity?** Although a response plan may be designed with a particular incident in mind, an actual emergency may be smaller or larger, simpler or more complex than planned. Though this question is related to the previous two, it is important to specifically examine how the problems that might arise affect performance at incidents requiring different levels of response performance. While some problems might not matter at all in relatively small or simple incidents, their effects might be decisive in relatively large or complex ones. As a result, for a given response system, performance would likely

be more predictable, and confidence in the system therefore higher, for incidents with more modest performance requirements.

Looking at response operations in this way—asking what might go wrong and how it might affect performance—is essentially asking "How *reliable* is the response system as designed?" The concept of reliability is more familiar in the context of technical systems: For example, a reliable automobile is one that doesn't suffer repeated breakdowns that make it hard to know whether it will get you where you need to go. But the same concepts can be applied to emergency response systems.[10] Reliable emergency response systems will be those in which only a few things might go wrong, the likelihood of those problems occurring is low, and their impacts on performance will be modest. Both the public and policymakers can reasonably have high confidence that those systems will perform well in the future. On the other hand, systems with more and more-serious potential problems should be viewed as less reliable, and confidence in them should be lower.

This approach to thinking about response performance more directly answers the fundamental question that the public and policymakers have about response performance. For a person who might be affected personally by a damaging incident or who represents those potential victims, the *theoretical* performance of the response system is much less important than its *likely expected* performance. After incidents in which performance did not meet expectations, the question "How do we know that the changes that have been made will produce better performance next time?" has been difficult to answer. If we could systematically assess and even measure a response system's reliability, we would be able to answer that question—by making a reasonable argument that changes and investments have addressed previous problems and, as a result, the system's reliability has increased.[11]

Beyond providing the public a "confidence assessment" of their response organizations, an understanding of both a system's overall reliability and the reasons why its reliability is high or low can inform a number of policy decisions as well. Measurement of response reliability could make significant contributions to setting priorities for future preparedness expenditures—e.g., investments that significantly increase the predictability of system performance at low cost would be very attractive. If such

[10] Assessment of the reliability of technical systems is the topic of an entire branch of engineering and a major element in design of complicated electronic and other technological systems. As we will discuss in Chapter Two, the analytic techniques and tools used in that field form the basis for the analysis discussed here for emergency response assessment. Because emergency preparedness and response systems are human systems, analysis of their reliability is necessarily more complex than for more well-defined technical systems. That complexity comes from the fact that they might be affected by more or less predictable problems (since the actions of individual people may be less predictable than the performance of technical components), but also from the fact that human systems can be more adaptable and flexible than technical systems.

[11] Reliability levels could also be an element of the performance objectives defined for emergency management programs as defined in standards such as NFPA, 2007, and EMAP, 2007.

opportunities exist, major improvements in future performance could be made at low cost. Measurement of response reliability could also be important for comparing preparedness in different areas and jurisdictions and understanding differences in planning, program costs, and potential outputs and outcomes of future response operations. Finally, being able to assess response reliability is critical for a well-informed public debate on emergency preparedness and response.

What is needed to actually assess the reliability of a response system? In short, we must—as systematically as possible—answer the three questions listed above about what might go wrong, how likely those problems might be, their impact, and how they would affect performance over the full range of incidents we expect our response systems to address. The approaches we apply to answer these questions need to be practical for emergency planners to use in the course of planning, so that the measurement and assessment process does not unduly compete with other preparedness activities for time and attention. The remainder of this document lays out the approach that we developed. Drawing on approaches and knowledge from disciplines ranging from emergency response practice to system engineering and statistical analysis, we describe a structured response reliability assessment that produces a number of results that can be used for preparedness planning and assessment in a variety of ways.

## About This Study and This Document

The concept of response reliability as a potentially useful addition to preparedness assessment was proposed by Brian Jackson in *The Problem of Measuring Emergency Preparedness: The Need for Assessing "Response Reliability" as Part of Homeland Security Planning* (2008). The goal of the research reported in the current document was to take that conceptual proposal and "prototype" the approach to more fully explore its utility and practicality. This work was part of a larger effort by the National Center for Risk and Economic Analysis of Terrorism Events (CREATE) for the Federal Emergency Management Agency, National Preparedness Directorate, National Preparedness Assessment Division. The study developed approaches using methods of risk analysis to support emergency preparedness planning and analysis.[12]

Reflecting that this study focused on developing a new methodology, the document is structured to build from a simple demonstration of the approach and its potential applications to a more complex and realistic example of its use. Chapter Two lays out the conceptual foundation for our response reliability analysis and works through a very simple example analysis to demonstrate the approach and the range of results it can produce.

---

[12] The CREATE effort examined four scenarios: an improvised explosive device attack on large public gathering, a chlorine release scenario (the focus of both the RAND work and analysis by others within the larger project team), an attack using a radiological dispersal device, and hurricane and flood events.

The remainder of the report examines a more realistic and relevant scenario, analyzing a response to a chlorine release. Chapter Three describes the schematic chlorine release scenario that served as the basis for that analysis and discusses the response parameters and options associated with that scenario. Chapter Four describes the system model of a response to the chlorine scenario, laying out the interactions between different elements of the response and how the overall effort meets the needs of victims of the incident. Chapter Five examines how the functioning of that system might break down, looking at failure modes for the various elements of the system.

Chapter Six discusses practical application of this approach to real preparedness problems. To illustrate how this approach could be used to analyze real-world datasets, we describe a prototype application of the analysis that uses the taxonomy of failure modes identified in Chapter Five to parse after-action reports to past response operations. Using this dataset, we demonstrate how this approach can be used to both integrate across and quantitatively analyze data on response performance.

Chapter Seven concludes with some observations on the relevance and value of the approach and exploration of a variety of ways these ideas could be drawn on in policy planning, preparedness evaluation, and the management of emergency preparedness systems. The appendixes provide some additional information related to the discussions in the text.

CHAPTER TWO

# Defining and Demonstrating Response Reliability Analysis

In this chapter, we introduce and demonstrate reliability analysis. We begin with a brief introduction to the way reliability analysis is done in the technical field, the source of the techniques and concepts we are drawing on and adapting for this work. Making the transition to preparedness, we then review how concepts similar to those we explore here have been applied to the analysis of emergency response systems.

The heart of the chapter is a demonstration of a reliability analysis on a highly simplified example response system. In this illustrative analysis, we show how answers to the three questions posed in Chapter One—How likely is it that individual problems that would affect response performance will occur? What effects would they have on the performance of the system? How do these effects vary for different incident sizes?—make it possible to systematically assess the reliability of that system.

Our goal in presenting this example case in detail is threefold. First, beyond just demonstrating our techniques in a more accessible way, this example case will also provide the opportunity to define some terms and approaches that we will apply in our subsequent examination of a chlorine response. Second, it provides a way to show the approximations and simplifications involved in different approaches to reliability analysis. Finally, a "fully worked" simple example makes it possible to explore—at the conclusion of this chapter—the full potential value of this technique for answering key preparedness policy questions.

## Defining the Analytical Process for Response Reliability Assessment

The concept of system reliability and its associated analytical approaches are used extensively in the analysis of technological systems to assess the likelihood that a specific piece of equipment—whether a key circuit board in a cellular phone or computer, a pumping system in a water treatment plant, or even a vehicle designed to take people or cargo into space—will perform over the period of time it is expected to function.[1]

[1] This is a restatement of the mathematical definition of *reliability* as measured for technical and other systems: "The probability that a system or component will function over some time period" (Ebeling, 1997, p. 23). We discuss how we have modified this definition for response reliability below.

Reliability and system safety analyses, drawing on techniques from engineering, statistics, risk analysis, human factors, and other fields, have been used to assess the functioning of systems in which both technology and people are involved, such as nuclear power plants and aircraft cockpit operations.[2]

In technical systems, analyses of system reliability are done for many reasons, including identifying and addressing problems during system design, understanding the likely future performance of the system under different conditions, and making cost-benefit judgments about specific alterations or repairs that might make the system's performance more predictable.[3] In considering emergency response systems, the ability to answer those same questions could directly contribute to identifying and correcting problems, providing the public with a measure of how the system might perform at future incidents or disasters, and informing cost-benefit or cost-effectiveness judgments regarding different potential investments in preparedness or response capacity.

### Component and System Reliability Analysis: An Overview

A common feature in reliability analyses is the recognition that systems and their components do not function perfectly and that events or circumstances will inevitably arise that affect their ability to do so.[4] Faults or failures arise because some part of the system does not work as expected, potentially hurting the performance of the system overall. Failures can happen because of the way the system is designed: Because building systems designed to function perfectly under all possible conditions is very costly, compromises that affect reliability are usually made, and the consequences of those compromises need to be understood. Or failures can be caused by some outside event or circumstance. Drawing the analogy to emergency response, a jurisdiction might rely on an antiquated public notification system that breaks when it is called on (an "internal design" failure), or the winds of an approaching hurricane might topple the

---

[2] Prominent examples of such applications are the work of Perrow (e.g., 1999) and Sagan (e.g., 1993), though there are a variety of other applications of these concepts to the functioning of organizations and the possibility of breakdowns in performance or failures with less technical focus than the examples cited here.

[3] For example, in the analysis of a technical system, engineers assessing potential problems that could reduce reliability will likely identify many changes or improvements that could be made to improve future system performance. Since available resources for system development are finite, however, judgments and trade-offs must be made about which issues are serious or important enough to modify the design and address (i.e., the cost of doing so is outweighed by their benefit) and which are not (or, more accurately, which potential problems might be better addressed by changes in operation and maintenance of the system, or repair if and when a failure occurs). Such time and budget constraints and their application in the cost-benefit judgments for what failure modes should be addressed are discussed in a variety of standard texts including Ebeling, 1997; Modarres et al., 1999; and Hecht, 2003).

[4] A variety of textbooks, technical standards documents, and other sources are available describing the techniques and concepts of reliability engineering and analysis far more comprehensively than the brief sketch provided here. Two that were drawn on in crafting this summary discussion are Modarres et al., 1999, and Ebeling, 1997.

antennas of an otherwise serviceable system, causing it to fail (an "externally triggered" failure).

As is likely clear already, understanding the reliability of a complicated system—in our case, an emergency response made up of many responders, technologies, and other ingredients—requires thinking about reliability at a number of different levels. The lowest level is the reliability of *individual components* of the system—the communications system supporting the response, the vehicles used to evacuate people, and so on—and what types of problems or events could affect the ability of individual components to function. The highest level is the reliability of the *system*, which takes into account how the interaction of the various components might make problems with a single component more or less important with respect to the functioning of the system overall. If a single component plays many roles within a more complicated system, then even a small reduction in its performance might have a disproportionate effect on the performance of the system overall. Conversely, if multiple backups exist that make it possible to compensate for a component's failure, problems that affect the component might affect system function only a little, if at all.

At both the individual component level and the system level, failures occur when some type of challenge or stress overwhelms the component or system's capacity to continue functioning as expected. Viewed this way, components and systems have some ability to function under adverse conditions—they have an inherent strength, endurance, or tolerance for particular types of stress—and failures only occur when the stress exceeds the threshold they can tolerate (see Modarres et al., 1999, pp. 2–4, for a more comprehensive discussion). Applying this view of failure to emergency response operations and performance, for a given response, some things that go wrong do not affect performance, or have effects that can be addressed with modest adjustments to response operations. But when the scale and effect of the problem reach a threshold, the system will not be able to compensate and performance will fall—in the language of reliability analysis, a failure will have occurred.

How is the occurrence of failures in potentially very complicated systems studied and understood? With enough information on a system and its components, one could, in principle, model its behavior and predict when failures would occur based on the understandings of chemistry, physics, engineering, and human behavior. If we had enough data about a communications system that was intended to support a response—its age, what sorts of things have gone wrong previously, how it has been used, and so on—then it might be theoretically possible to predict when it was going to break down. In practice, however, limits on what is known (and the costs associated with collecting so much data) mean that such projections are often difficult or impossible to make. As a result, analyses of real systems and projections of their reliability combine both understanding of how a system was built and how it functions with statistical approaches that model breakdowns from various mechanisms as well as from randomly occurring events (see Ebeling, 1997, pp. 4–5). For example, different approaches model failures as happening with constant rates (but with uncertainty associated with their exact occurrence) related to the amount of time that a component or system has been used or potential problems in its manufacture, among others.

To estimate the reliability of individual components and systems, analysts and engineers gather data on the occurrence of failures of different types over time, either during actual operation of a system or during testing processes over shorter time periods. Statistical methods are applied to these observed data to estimate the rates of failure (or the probabilities of individual failures occurring within a specific time period) at varying levels of confidence and precision.[5] In the absence of data, estimates can be made based on analogies to other technologies, or estimates can be made more broadly, based on technical expertise and experience. A number of different techniques can then be used to take estimates of the reliability of individual components and combine them to build estimates of the performance of the complex system assembled from those components. These techniques, which vary both in their difficulty and the types of approximations made, enable different types of system reliability assessments for different design, evaluation, and analytical purposes. It is this palette of techniques that we drew from to craft an approach for evaluating the reliability of emergency response systems for preparedness assessment.

### Adapting Reliability Analysis Techniques to the Evaluation of Emergency Response Systems

Using reliability analysis as the basis for preparedness assessment requires translation of techniques developed in another field so they can be applied to a different problem set. Although we are not aware of examples that do this in the general way we have attempted to do here, concepts from reliability analysis and related fields have previously been used in some analyses and evaluations of emergency preparedness.

Emergency response operations have been a topic of some interest in the operations research field for many years.[6] In these analyses, the major focus has been on issues related to "everyday performance" of response systems—e.g., the response of fire and emergency medical services (EMS) to the types of emergency events that happen regularly. Many of these analyses use metrics that either explicitly or implicitly reflect concepts of reliable response system operation, though other measures, such as efficiency and cost minimization, are also prominent. Many of these analyses focus on problems such as the placement of response base locations and routing vehicles, using quantitative modeling and classical operations research techniques (Simpson and Hancock, 2009). Measures relating to the reliability concepts of interest here include

---

5 See discussion in Modarres et al., 1999, or Ebeling, 1997, for a methodological description of the techniques involved.

6 See Simpson and Hancock, 2009, for a review of past operations research on everyday emergency response and Altay and Green, 2006, for a similar look at its (comparatively limited) application to disaster response operations.

the probability that response units will reach locations of response need within threshold time periods.[7]

There are also real-world examples of this sort of measure being applied to response organization planning and evaluation. NFPA standard 1710 (NFPA, 2001) on the performance of career fire service organizations has an explicit reliability requirement embedded in it, requiring that 90 percent or more deployments to fires occur within defined response times. A publicly available example of reporting from a department based on this standard that we identified is the *Standards of Coverage* report produced by the Ashland, Oregon, Fire and Rescue Department. In addition to reporting percentages for achieving threshold response times, it also includes an assessment of response reliability, which it defined as the "probability that the required amount of staffing and apparatus will be available when a fire or emergency call is received"[8] (Ashland Fire and Rescue, 2009, p. 39). Similar logic (if not explicit measures) is also part of other key standards for emergency management planning (e.g., NFPA, 2007; EMAP 2007). For example, in its discussion of resource management, NFPA standard 1600 includes a requirement for "contingency planning for shortfalls of resources" (NFPA, 2007)—essentially, planning to ensure reliable operations in spite of events that could cut the resources available for response.

Explicit application of these sort of approaches and concepts to the type of temporary response organizations and activities that are involved in large-scale incident operations—up to and including disaster-scale incidents—is much more rare. A recent review (Altay and Green, 2006) emphasized the relative scarcity of the literature on these topics, and Simpson and Hancock (2009) called out performance measurement as an area that is particularly underdeveloped. In a search of the literature performed for this study, we found some examples of analyses of large-scale incidents and response operations that incorporated these concepts. Examples from analyses of evacuation planning were prominent,[9] and we identified additional examples examining reliability issues associated with the incident command system used for disaster response management (Bigley and Roberts, 2001); the reliability of emergency systems, with a particular focus on inter-organizational cooperation (Kanno and Furuta, n.d.); vehicle routing post-disaster (Jotshi et al., 2009); and the design of locations for supply depots (Rawls and Turnquist, 2010) or such activities as mass antibiotic dispensing (Hupert et al., 2002; Lee et al., 2006; Nelson et al., 2008). A more analogous analysis to what we

---

7 To provide a context for our work, we searched for relevant papers from this literature that discussed measures that included reliability concepts. Selected examples that include probabilistic methods for describing response performance or other reliability related measures include Kolesar et al., 1975; Chelst and Jarvis, 1979; Revelle and Hogan, 1989; Ball and Lin, 1993; Beraldi et al., 2004; Fry et al., 2006; Dausey et al., 2008; Iannoni et al., 2008; Silva and Serra, 2008; Afshartous et al., 2009; Pal and Bose, 2009; Peeta, et al., 2010; and Sorensen and Church, 2010.

8 This is similar to, though not identical, to the definition of response reliability we will use below.

9 For example, Han et al., 2007; Georgiadou et al., 2007; Stepanov and Smith, 2009; Chen and Zhan, 2008.

describe is Arboleda et al. (2007), which uses a system dynamics model to assess the vulnerability of the functioning of a health care facility to breakdowns of key infrastructures on which it depends. Peeta et al. (2010) address how damage to transportation infrastructure during a disaster could hurt response operations and how additional investment could reduce those effects. Other analyses of specific types of incidents have used similar methods to examine activities such as bioterrorism response operations, though they have used metrics that differ from those of interest to us here.[10,11]

The relative scarcity of such analyses perhaps should not be surprising, as making the jump from assessing the reliability of response to everyday emergencies to evaluating responses to larger and rarer incidents is not a trivial exercise. For everyday operations, performance data collected on each call can provide the basis for assessment (see Ashland Fire and Rescue, 2009, for an example). For analyses like those focused on the placement of fire stations, the fact that the geography of the system can be well defined makes it possible to use models to simulate traffic flows and other factors that might affect the transit time of a response unit from its home station to a call location. Though potentially still complicated analyses to perform, the ability to define the problem very specifically makes it possible to do highly quantitative analyses of how changes in the system or the circumstances in which it operates will affect measures such as average response time and the variance around that average.

But for assessing preparedness for more unusual or larger events, day-to-day experience doesn't provide all the information needed. Though large-scale incident response operations are based on the same management structures and processes used in everyday response, in many ways the response to a large-scale event is custom-built for the incident at hand. Though a fire department or other agency will often operate as a single organization, large-scale events are usually multi-agency affairs that require unified operations, may involve multiple levels of government, and are different in other important ways from everyday activities. There are often more things that can go wrong and different failure modes that might hurt response performance.[12] This makes applying reliability analysis concepts to such response operations a messier analytical problem. The response systems that are implemented for large-scale incidents can be complex, and they are also human systems with the associated strengths and weaknesses. Simpson and Hancock (2009, p. S134–S135) called out such systems as

---

[10] For example, Kaplan and Walden, 2007; Wein et al., 2003; Wein et al., 2002. O'Reilly et al., 2005, and Conrad et al., 2002, appear to address emergency response in larger-scale disasters, but focus on disaster effects on everyday response operations.

[11] There are also examples of this type of analysis being used for other types of large-scale expeditionary activities. For an example focusing on military operations, see Kelley, 2004.

[12] Reviews of problems that occur with great regularity in large-scale response operations are readily available in the practitioner and disaster response literature. Recent examples include Donahue and Tuohy, 2006, and Larson et al., 2004.

requiring a "softer" operations research approach, since they cannot be as well defined as problems related to everyday operations.

Furthermore, if an analytical approach is to contribute most broadly to response planning and evaluation, it must be simple enough to be replicated and applied. As a result, there is a premium on keeping the methodology as simple as possible. If a reliability analysis requires that the full details of every possible incident be simulated and the characteristics of the system specified down to the level of fire station locations and road systems, it will be impractical for many desirable applications.

In applying reliability analysis to emergency response planning, we drew on a set of techniques from the reliability analysis and system engineering fields that support systematic assessment of a system's reliability characteristics at varying degrees of precision or approximation. Those techniques are failure mode and effects analysis (FMEA) and the related failure mode, effects, and criticality analysis (FMECA).[13] For our work, we found FMECA attractive because the approximations that are made allow each failure mode to be treated independently, somewhat simplifying use of the technique for assessing a complex response system.[14]

These techniques are generally applied in the design of systems as a way of identifying failure modes and their effects—and, if the result is viewed as unacceptable, to return to the design phase and address the reliability problems.[15] To the extent that the goal of preparedness evaluation is making policy changes or allocating resources to improve future performance, the end application of our use of the techniques is quite similar to the iteration on the design of a technical system before production.

FMECA analyses involve a structured set of four steps for describing a system and then identifying and analyzing the ways it could fail. We have adapted these steps to response systems, to address the three core questions laid out in Chapter One. We have diagrammed our version of the four steps Figure 2.1, and we will use this figure as a visual map in the remainder of the sample analysis in this chapter and in subsequent chapters, to orient the reader to what portion of the analysis is being discussed.

1. **Define and Map the System.** Understanding what might go wrong in a system requires knowing how it is put together. Laying out the different functions (in

---

[13] A variety of sources describe these techniques and their application. For example, Ebeling, 1997, pp. 166–173; Modarres et al., 1999, pp. 262–267; DoD, 1980; U.S. Army, 2006; U.S. Nuclear Regulatory Commission, 1981; FAA, 2000. FMECA is only one of a number of methods in reliability analysis for combining information on the incidence and consequence of failures to an overall assessment of a system.

[14] Note that this treatment—by treating individual failures on their own—does not consider the possibility of many small (capability reducing in our categories) failures adding to the point where they cause system collapse (response termination.) Rudolph and Repenning, 2002, suggest such a model where individual failures increase the stress on a system to the point where it breaks down.

[15] *NFPA-1600*, the emergency management standard, cites FMEA as a technique in the context of this sort of assessment—how hazards could create failures at facilities or in systems society depends on as part of framing response requirements (2007, p. 11).

**Figure 2.1**
**The Four Steps of Response Reliability Analysis**

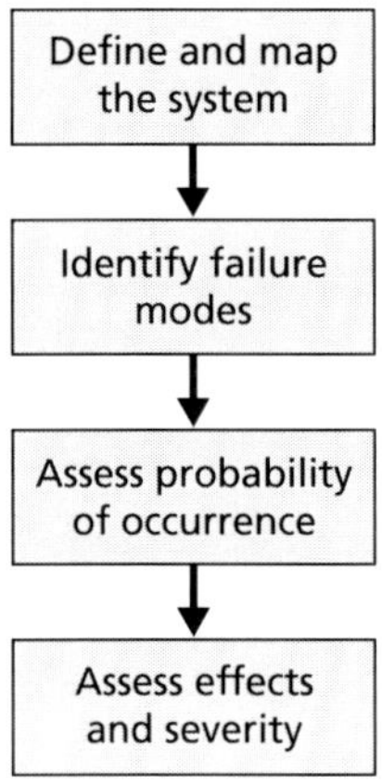


RAND *MG994-2.1*

the case of response operation) that must be performed, the varied organizations and agencies involved in performing them, and how their activities and capabilities link together defines the structure and bounds of the analysis. This is generally done as a "block diagram" showing system elements and their linkages. As part of system definition, what it means for each part of the system to work well and how reductions in performance would translate to poor outcomes need to be determined.

2. **Identify Failure Modes.** Failure modes are defined as "the observable manners in which a component fails" (Ebeling, 1997, p. 168), which in this case would be the ways that performance of different parts of the response system would break down. Identifying failure modes includes systematically inventorying what might go wrong in each part of the system. The potential timing of failures is also important: Some failures might occur as a response operation was being initiated, while others might occur at later points in the response. Failures may be due to random events (e.g., equipment failures even though all appropriate maintenance had been done), have a clear human cause (e.g., maintenance had not been done), or be caused by external events (e.g., the incident damaged the vehicles prior to deployment).
3. **Assess the Probability of Occurrence of Different Failure Modes.** Given many things that could hurt the functioning of a system, one differentiator among them is how likely they are to happen. The probability that a specific failure will occur during a response operation could be estimated a number of different ways; for example, the estimate might be based on real-world data on the occurrence of failures in past responses, or it might be elicited from subject-matter experts. Each method has its own strengths and weaknesses with respect to the types of approximations or biases involved, but practical or other factors

could drive the choice among different approaches. FMECA can be applied using estimates of probability.[16] Depending on how failure modes have been defined, some calculation may be involved in determining the probability of a specific mode.

4. **Assess the Failure Mode Effects and Their Severity.** The other differentiators among failure modes are their effects and the severity of those effects. In FMECA, this assessment is done at the system level, by asking "What is the effect of the failure mode's occurrence on system performance?" Failure modes can have a variety of effects, ranging from complete failure of the overall system to essentially no effect at all. Intermediate between such extremes are failures that might degrade but not terminate system functioning.[17] In considering the effect of failures on emergency response operations, a simple analysis could be viewed as involving two classes of failure:
   a. **Response-termination failures:** Failures that would stop a response operation entirely, equivalent to the most serious class of failures in traditional FMECA analyses.
   b. **Capability-reduction failures:** Failures that make a response operation less effective but do not halt response (e.g., an event that reduces the number of victims a hospital could accept after an incident). Capability-reduction failures include failures that cause a reduction in system performance either directly or via their effects on other response functions—for example, difficulties implementing incident command could have an effect on many other response activities. How severe the effects of such failures would be for system performance would be driven by how large their impacts were relative to the requirements at the response operation.

   This judgment about severity addresses the third question posed in Chapter One regarding the effects of failure modes—since an identical failure might be much more severe in terms of system performance at a large or otherwise demanding incident (where the response system was stretched close to its limit) than at a small or more straightforward one. The effects of a particular failure mode may be estimated in terms of a percentage loss of response performance (e.g., loss of some number of vehicles cuts evacuation capacity by some percentage) or absolute loss of capability. As with probability values, estimates might

---

[16] Descriptions of the technique include a standardized table of probability levels as follows: (a) frequent—probability of failure greater than or equal to 20 percent; (b) probable—probability from 10 percent up to 20 percent; (c) occasional—probability from 1 percent up to 10 percent; (d) remote—probability from 0.1 percent up to 1 percent; (e) extremely unlikely—probability less than 0.1 percent (adapted from Ebeling, 1997, p. 170).

[17] In standard descriptions of FMECA, four severity classes are used: (I) "Catastrophic—Significant system failure occurs that can result in injury, loss of life, or major damage," (II) "Critical—Complete loss of system occurs; performance is unacceptable," (III) "Marginal—System is degraded, with partial loss of performance," and (IV) "Negligible—Minor failure occurs, with no effect on acceptable system performance" (Ebeling, 1997, p. 169).

be made from past experience with large-scale responses (e.g., how disruptive to operations specific failures actually were) or through processes of expert elicitation or projection of the *likely* effects of particular events.[18]

For performing these four analytic steps for emergency response systems, existing response standards, doctrine, and other sources provide a foundation for both framing and performing assessment. The first analytic step, defining the architecture and characteristics of the response system, is consistent with how emergency management efforts are defined in existing standards and doctrinal documents. For example, in the Emergency Management Accreditation Program (EMAP) standard, an emergency management program is defined as a "jurisdiction-wide system that provides for management and coordination of prevention, mitigation, preparedness, response and recovery activities for all hazards. The system encompasses all organizations, agencies, departments, entities and individuals responsible for emergency management and homeland security functions" (EMAP, 2007, p. 1). Our conception of an emergency response system parallels this definition, though we focus on the preparedness and response activities of all the entities involved. Other sources, including existing scholarship, practitioner expertise, and response doctrine (e.g., national-level documents such as the TCL [DHS, 2007b] and *Universal Task List* [DHS, 2007a]) could also contribute to defining the response system and to later components of the FMECA analysis as well.[19]

For assessing a specific area's preparedness, its preparedness plans and associated procedures would provide much of the information needed for the other analytic steps. Indeed, elements of this process (e.g., identifying what might go wrong with different parts of a response operation) represent elements of good practice in planning for response operations. The planning process requirements defined in key response standards (e.g., EMAP, 2007, or NFPA, 2007) include the identification of failure modes and the implementation of measures to address them.[20]

In applying these steps, however, analysis must also take into account that response systems are human systems and are therefore more flexible than the technical systems that are the usual focus of system reliability analyses. The properties of a

[18] In FMECA, a criticality index for each failure mode is calculated based on its probability of incidence and, combined with the severity of the outcome, used to prioritize which failure modes should get the most attention. The index combines information about the importance of a failure mode (i.e., for a given element of the system, the number of different failures caused by a single failure mode type), the probability that failure of the component will produce the effect on system performance (i.e., if it does fail, is a catastrophic outcome probable, possible, or unlikely?), and relevant failure rate and time variables (see Ebeling, 1997, p. 170).

[19] As we will discuss in later chapters, these sources were useful in our analysis in identifying potential failure modes and characterizing their potential effects.

[20] For example, some of the requirements for leadership succession directly address failure modes involving the unavailability or loss of key commanders, and the exercise and evaluation/corrective action program elements embedded in these standards are designed to identify and correct failure modes over time.

technical system are generally set at design and manufacturing, and so modifications and adjustments made in the course of operations are limited to elements that have already been built in (e.g., the presence of redundant or backup system elements). Human systems have the potential to be much more flexible, adjusting to circumstances as they appear—meaning that a failure mode analysis of a response system must take into account not only the events that are possible failures, but also the steps that might be taken "on the fly" to adapt to their occurrence, and what the residual effect on response performance would be. Some response systems (those with more-comprehensive plans that hedge against more possible failure modes, with more highly trained and flexible leadership, and so on) will be more flexible and adaptive than others. Human systems also have other vulnerabilities that technical ones do not. For response operations that involve many different agencies, the personal relationships that exist between leaders are a key element of the "wiring" that links different parts of a response operation together. In a technical system, such wiring might be essentially static and easily analyzed. In a human system, the strength of connections will change over time with personnel rotations, multi-agency exercises and coordination, and so on. These characteristics make these systems more difficult to analyze than technical systems; doing so might require more approximation and estimation, but the analytic tools and approaches are useful nonetheless.

Analysis of response systems must also take into account the fact that such systems are much more dynamic and mutable than a piece of electronic equipment or a nuclear power plant, for which the wiring and subsystems can be laid out in a static circuit diagram. For large-scale events in particular, the nature of the response system will almost certainly change over the course of the response:

- In the initial phases of a disaster response, most of the activity will be local, and the local response system (perhaps including only agencies and organizations in and near the affected area) will likely be operating at, near, or even beyond the limits of its designed capacity. In such a situation, failure modes arising from resource scarcity will likely be critical and, as a result, adaptations and changes in what the system is trying to accomplish and how it is trying to do so will likely be made, in an effort to help the most people as much as possible.
- As additional aid arrives, from regional or federal sources, *the response system itself will change*—other organizations and their capabilities will be plugged into operations, additional supplies will arrive, and so on. As more and more organizations become involved, failure modes associated with resource scarcity will become less important (the core rationale behind such mutual aid models and multi-agency responses), but failure modes associated with interagency coordination and integration could become much more important, simply because more organizations that do not usually work together have become involved.

Failure mode analysis of the response to large-scale events must reflect the fact that the response system itself can evolve over time. As the number of organizations and resources available and involved increases, the system will transition from one state to another—gaining new capabilities that might address failure modes that limited what it could do in its smaller state, but also likely taking on additional failure modes (some of them associated with the period of transition) that can affect its performance.

## A Simplified Response Example for Defining and Illustrating Response Reliability

To explain the various steps involved in applying methods of response reliability to preparedness, it is easiest to walk through a sample analysis. In this example, we demonstrate the individual steps of the process and explore some of the insights regarding preparedness that can be gained at each step. Using a notional example makes it possible not only to illustrate each element of the analysis process clearly (i.e., without complexities of data and mathematics getting in the way of explaining the thought process and concepts behind it), but also to illustrate the full potential of the analysis in an ideal case—specifically, a case where *quantitative* measures of response reliability can be developed.

To focus on the analytical process rather than the details of the example, we have constructed a highly stylized and (over)simplified response case:

> An incident has occurred in which a particular medical treatment has to be delivered to a subset of the public within a fixed period of time. The local response organization has 300 operational responders, each of whom is qualified to deliver the treatment. In preparing for this incident, the organization has staged all the supplies that are needed at a central location in its jurisdiction and intends to manage response operations from that location. To initiate response, all necessary responders are expected to assemble at the central dispatch/staging location and deploy from there. To deliver the treatment to a member of the public, a responder has to have mustered at the staging location, collected the supplies necessary for all of the treatments during the response, been told where a member of the public is that needs assistance, and traveled to that location. Once the treatment is delivered, the responder can be retasked to help the next victim.

Figure 2.2 presents an illustrative cartoon of this example response operation.

Based on the assumptions in the area's preparedness planning, each responder is expected to be able to treat an average of approximately five people during the available window of time. Assuming no random variation in the rates of treatment, this would correspond to an approximate maximum planned response capacity to treat 1,500 people in the time available. That maximum response capacity (which we refer to as

**Figure 2.2**
**A Simplified Response Operation**

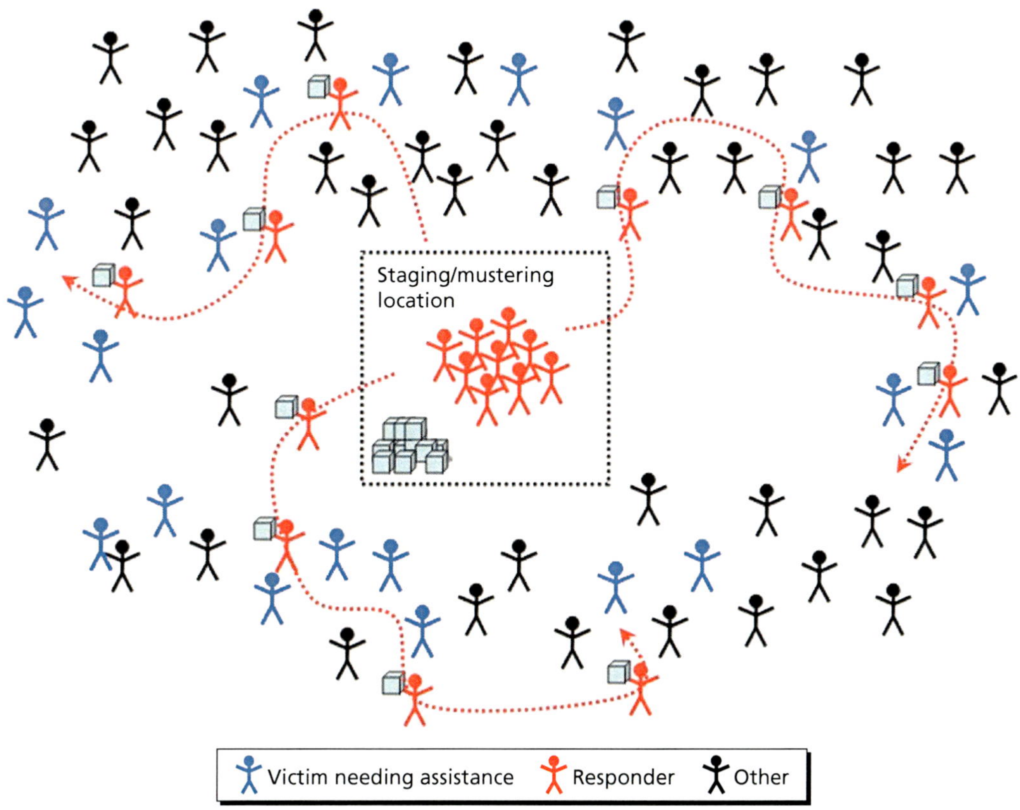


NOTE: Response defined for illustrative case as delivering a treatment (with supply package) to victim.
RAND *MG994-2.2*

$RC_{max}$ in later discussion) represents the highest level at which the system would be expected to perform, assuming its plans can be implemented as written[21] and nothing goes wrong. But, events likely will occur that reduce the response system's ability to perform. Accounting for those failure modes and their effect on response performance is the crux of our analysis.

Before walking through the analytic steps for this simple system, we note that having this specific example makes it possible to better define some of the abstract concepts introduced in the previous section. Briefly,

---

[21] Identifying 1,500 as the maximum expected response capacity, as we have done here, simplifies response operations down to a deterministic system in which the average response rate represents an accurate estimate of performance at any given incident. In reality, there will be random variation in response performance (e.g., differences in average response rates) that will mean that 1,500 would be more realistically viewed as the center of a distribution of possible outputs at the upper end of the response system's capacity. Later in the chapter, we will include this sort of random variation as we analyze this example.

- **Different failure types can be defined by how they affect the system's ability to produce response outputs.** In this simple case, the output is the number of "treated patients." Response-termination failures stop all activity, halting any treatment. For example, a major problem at the staging area that let no responders exit across the black dotted line in Figure 2.1 would be such a failure. Capability-reduction failures would reduce the number of people that could be treated in the time available for response. For example, some staff might be unavailable (fewer red responders), or events might slow down responders' progress in assisting victims (reducing the average number of patients that each responder treats).
- **To be useful, this analysis needs to address reliability not just for the largest and most demanding incidents, but for smaller and less demanding ones as well.** Though we have talked so far about the maximum response capacity of this simple system,[22] we must recognize that a real incident could fall in a broad range of size, scale, or complexity—and it is likely that incidents below a system's $RC_{max}$ will arise more frequently. Clearly, incidents that were larger than the system's maximum planned performance level would be a problem. But for incidents well below the $RC_{max}$, a response system would reasonably be expected to be able to perform more reliably. Failures modes that cut into total response capacity will be tolerable for a system if the response *requirements* are significantly less than its $RC_{max}$. Slack capacity in the system makes it possible to absorb some faults before overall system performance drops below an acceptable level—for this simple case, the minimum acceptable performance is defined as the ability to deliver treatment to all those that need it at any arbitrarily sized incident.
- ***When* a failure occurs during a response can greatly affect its severity.** Failures that happen at the very beginning of an operation will affect it in its entirety—and if those failures cause termination of the response operation, they will affect performance at (and therefore reduce system reliability for) incidents of any size. On the other hand, a capability-reducing failure that occurs late in a response will have much less effect on output than if it occurred early. Indeed, for small incidents, response operations are likely to be shorter overall (with fewer people needing assistance), narrowing the time window for some failures to happen at all.

With these concepts defined, it is now possible to more clearly state how we define response reliability. Given a specific response system designed to respond to incidents up to and including some maximum assumed scale, scope, or complexity and, therefore, response performance requirement, we define the response reliability of that system as follows:

[22] In our simple example, the response performance required of the system is the total number of victims needing treatment—the goal is to treat everyone. In this example, that number would track directly with the size of the incident, and so the more specific terminology of *incident size* (i.e., number of victims) and the more general response performance required are interchangeable. For other response capabilities and incident types, this would not necessarily be the case.

> The probability that the response system (defined by a set of plans, resources, authorities, agencies, and their associated human resources) will be able to deliver at or above a given level of capability or performance at an actual emergency incident.[23]

For a single system, response reliability will therefore be different for incidents of different sizes—since smaller or simpler incidents are "easier" than larger or more complex ones. Our example system designed to provide up to 1,500 treatments immediately after an incident would have a reliability value that starts at 100 percent for an incident of size zero (since any system is perfectly reliable by definition if nothing is required of it), with decreasing reliability as incident size increases. Treating 1,500 patients as a hard upper bound on performance, reliability would drop to zero for a requirement to treat 1,501 patients, since the system was not designed to perform at or above that level.[24]

To illustrate how a response reliability assessment is put together, and how an FMECA-type analytic process can help to assess how system reliability changes between these two extremes of performance, the following sections demonstrate the four steps described previously for this simple response system.

### Step One: Define and Map the System

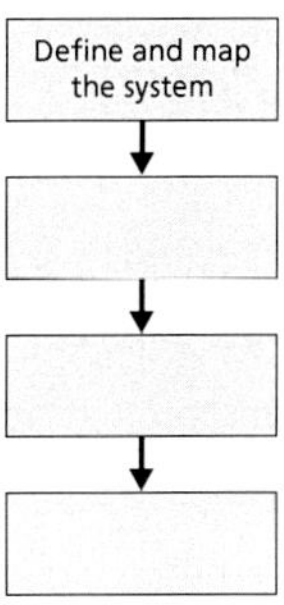


The first step of analysis is to lay out the different steps and functions that need to be performed in the response operation. For our simple system, the operation is run at one place and all response resources are co-located—and the only response task is deployment of supplied responders from that location to find and treat patients. To do so, the system has to perform the following tasks:

- **Initiate response operations**—which we have modeled as including four basic functions: receiving notification of the incident, dispatching staff to the staging location, those staff successfully getting to that location, and initiating incident management.

---

[23] Our definition of response reliability echoes Ashland Fire and Rescue's ("probability that the required amount of staffing and apparatus will be available when a fire or emergency call is received"), though ours is framed somewhat more broadly.

Comparing our definition of system reliability with that from the reliability engineering literature, the most important difference is what characteristic is being assessed. For a mechanical device, the measure frequently of most interest is time before failure—so reliability is defined in terms of the probability that a failure will not occur over a time period of interest (e.g., Ebeling, 1997, p. 5). In our case, we have taken a single response operation as the "time" of interest (accepting that response lengths will vary from incident to incident) and framed reliability in terms of the probability of achieving a specific response performance level.

[24] In reality, when the performance of a response system is understood to include variability in performance—i.e., that a system designed to an overall maximum capacity of 1,500 might perform somewhat above or somewhat below that level—then reliability would not necessarily drop to zero at the $RC_{max}$. How far above that level it might perform would depend on how wide the expected variance in its performance was at any given incident.

Figure 2.3
Basic System Diagram for Our Example Response Activity

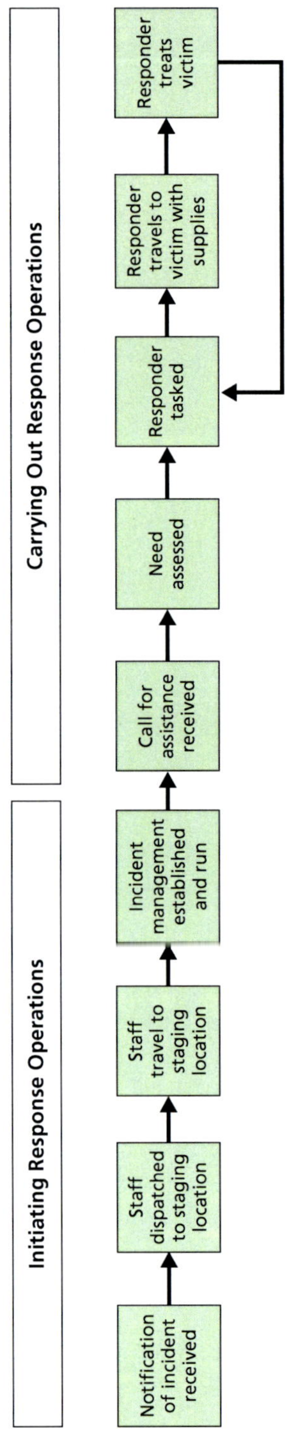


RAND *MG994-2.3*

- **Carry out response operations**—which we have modeled as involving receiving a call for assistance (which is the only way the victims become known to the responders to keep this example simple), assessing the need of the caller, tasking a responder (with supplies), travel of that responder to the victim, and treatment. The responder is then available again for retasking.

The map of the response is illustrated in Figure 2.3. Since the goal of this response operation is treatment of victims within a time window available, the system has two main "measures of merit"—indicators that it is performing well. The first measure of merit is the response organizations' ability to perform each function at all (i.e., avoidance of any response-termination failures). The second measure of merit is the rate at which the response organizations (as part of the overall response system) can treat victims. Given the belief that the system can serve 1,500 people within the time window available for response, assumptions about how long each of the tasks takes to perform are components of the response plan. Delays, mistakes, or other failure modes that result in tasks taking longer than assumed would cut into system performance.

### Step Two: Identify Failure Modes

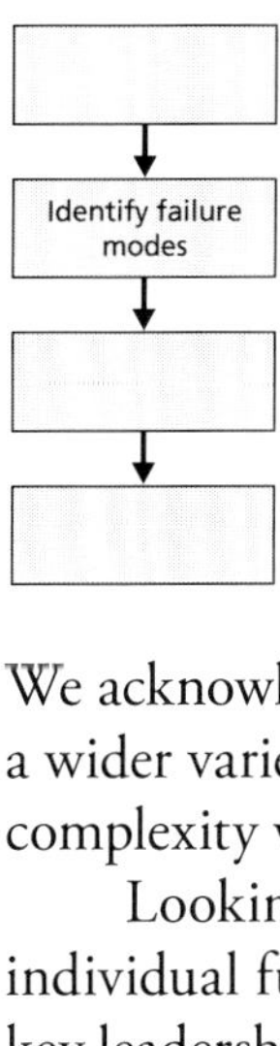


The core of a failure mode analysis is articulating what might go wrong that would affect the functioning of the response system. Identifying failure modes involves systematically thinking through each part of the system to determine what events would hurt performance. However, since a single failure mode could affect multiple functions within the response system, how failure modes potentially map to multiple functions must be considered as well. For the purpose of this discussion, we will limit this simple example to ten different failure modes that illustrate different types of failures and their effects. These are listed in Table 2.1. We acknowledge that even a simple system like that pictured in Figure 2.1 could have a wider variety of things that might go wrong, but the goal in this short list is to limit complexity while illustrating the process.

Looking at this list, it is immediately clear that some failure modes are specific to individual functions (e.g., 2, 10), whereas others are more general (e.g., 3—absence of key leadership; the attendant incident management disruption could affect many functions). Figure 2.4 illustrates how we have mapped the failure modes to the pieces of the exemplary response model. In this mapping, we have included some links that are weaker than others. For example, the effect of disrupted responder communications on the ability of responders to travel to victims would presumably manifest only if a victim could not be found and further information needed to be sought from incident command—this is a relatively weak linkage compared with the effect of communications breakdowns on the ability to dispatch responders in the first place. But in the interest of simplicity in discussion, we have not included every possible linkage.

**Table 2.1**
**Ten Failure Modes Associated with Our Example Response System**

| Number | Failure Mode |
|---|---|
| 1 | Response communications systems suffer intermittent breakdowns. |
| 2 | Calls from members of the public not needing assistance ("worried well") overwhelm systems. |
| 3 | A key member of the response leadership is traveling, disrupting the functioning of incident management. |
| 4 | Some responders needed to implement the plan are unavailable. |
| 5 | Supplies that were assumed to be at the staging area had been used and not replaced. |
| 6 | Members of the public cause physical disruption at the staging area. |
| 7 | Logistics management at the staging area is disrupted. |
| 8 | Higher-than-expected traffic in the area slows all travel and transportation. |
| 9 | Higher-than-expected breakdowns of response vehicles occur during operations. |
| 10 | Treatment of individuals takes longer than expected because responder training on process had not been done recently. |

### Step Three: Assess the Probability of Occurrence of Different Failure Modes

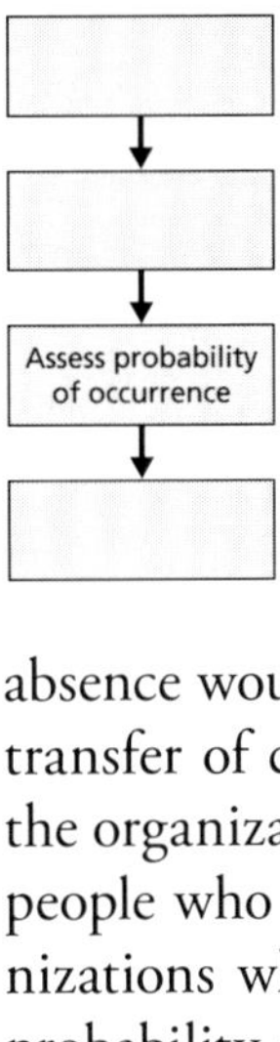


Having identified what might go wrong with the various elements of the response system, the next step is to assess their probability. Other factors equal, failure modes that are very likely are of more concern than those that are rare. Estimates of the likelihood of different failure modes are inherently specific to an individual jurisdiction, its circumstances, and its plans. Taking the absence of a key member of response leadership as an example, a jurisdiction in which the leadership rarely traveled would clearly rate this mode as low probability. In a jurisdiction where the leadership traveled frequently, the question would be whether their absence would disrupt incident management. If extensive planning had been done for transfer of command and cross-training had been done among ranking members of the organization, the probability might still be rated as low (since there would be more people who could fill the key positions in incident management). However, for organizations where individual members of the leadership were more indispensable, the probability of this failure mode would be ranked higher. Our description of FMECA above included an anchored scale for qualitatively ranking probability of occurrence of different failures linked to probability ranges, based on the way this technique is used in engineering analyses. For the purposes of this example, we will define a somewhat simpler scale: High (for failure modes viewed as having a 5 to 10 percent chance of occurring at any given response);[25] Medium (for a 2 to 5 percent chance); Low (for a

25 Having the "high" range end at a 10 percent chance of incidence could be too low for certain types of failures, or for particular ways of framing an analysis. It is done here in the interest of simplicity; more complex formulations will be discussed later.

**Figure 2.4**
**Mapping Exemplary Failure Modes to Model Response Functions**

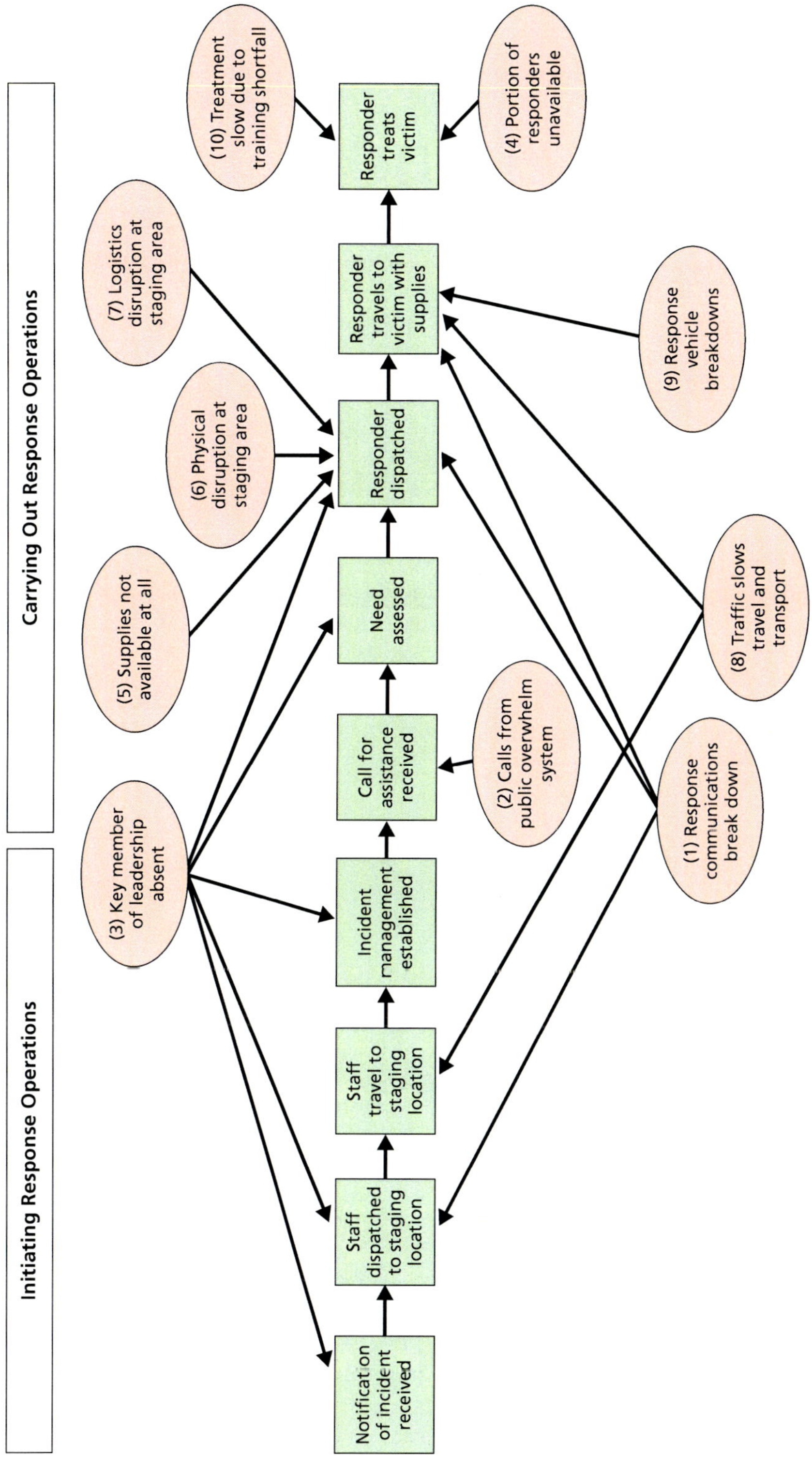


RAND *MG994-2.4*

1 to 2 percent chance); and Remote (for a less than 1 percent chance). In this case, the numbers are intended to simply provide a scale for what we mean by the different categories, although we will explore the value of assigning numerical values in later steps of the analysis.

In an actual analysis, these probabilities could be assigned based on data describing what problems occurred in past responses in the jurisdiction[26] or simply by practitioners making estimates of the relative likelihoods based on their past experience in the response organization (ideally through a structured elicitation process).[27] Continuing with our simple example, in Table 2.2 we have assigned probability categories to the various example failure modes. We chose to assign values across the range, but attempted to select ones that would represent a credible example of a realistic response system. We will continue to build on this base table in the later steps of this example response reliability analysis.

**Table 2.2**
**Notional Probability Levels for Example Failure Modes**

| Number | Failure Mode | Probability of Occurrence |
|---|---|---|
| 1 | Response communications systems suffer intermittent breakdowns. | Low |
| 2 | Calls from members of the public not needing assistance ("worried well") overwhelm systems. | High |
| 3 | A key member of response leadership is traveling, disrupting the functioning of incident management. | Medium |
| 4 | Some responders needed to implement the plan are unavailable. | High |
| 5 | Supplies that were assumed to be at the staging area had been used and not replaced. | Remote |
| 6 | Members of the public cause physical disruption at the staging area. | Low |
| 7 | Logistics management at the staging area is disrupted. | High |
| 8 | Higher-than-expected traffic in the area slows all travel and transportation. | High |
| 9 | Higher-than-expected breakdowns of response vehicles occur during operations. | Low |
| 10 | Treatment of individuals takes longer than expected because responder training on process had not been done recently. | Medium |

26 Failure rates and probabilities are estimated from observed data using statistical techniques (see Ebeling, 1997, or Modarres et al., 1999, for a discussion).

27 We will discuss an example of such an analysis using real data in subsequent chapters. The various ways of making these estimates each have strengths and weaknesses with respect to the quality of their results and the practicality/ease of their application.

### Step Four: Assess the Failure Mode Effects and Their Severity

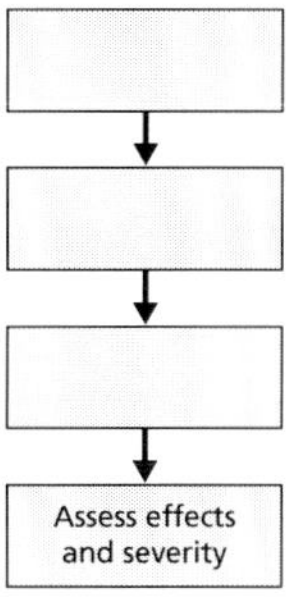


Even with some estimate of the likelihood of particular failure modes occurring, making a judgment about their relative importance still requires assessing their effects and severity. At a very basic level, the first part of assessing a failure mode's effects is cataloging how many different places at which it will potentially affect response performance. Table 2.3 does that, taking the mapping of which parts of the model would be affected by different failure modes (Figure 2.3) and adding that as an additional column in our growing response reliability analysis table. From even this general accounting, some initial observations can be made. Though shown as low probability in our example, communications failures affect the functioning of three different functions, whereas the more probable event

**Table 2.3**
**Mapping of Example Failure Modes to Functions Affected**

| Number | Failure Mode | Probability of Occurrence | Functions Affected |
|---|---|---|---|
| 1 | Response communications systems suffer intermittent breakdowns. | Low | • Staff dispatched to staging location<br>• Responder tasked<br>• Responder travels to victim with supplies |
| 2 | Calls from members of the public not needing assistance ("worried well") overwhelm systems. | High | • Call for assistance received |
| 3 | A key member of the response leadership is traveling, disrupting the functioning of incident management. | Medium | • Notification of incident received<br>• Staff dispatched to staging location<br>• Incident management established<br>• Need assessed<br>• Responder tasked |
| 4 | Some responders needed to implement the plan are unavailable. | High | • Responder treats victim |
| 5 | Supplies that were assumed to be at the staging area had been used and not replaced. | Remote | • Responder tasked |
| 6 | Members of the public cause physical disruption at the staging are. | Low | • Responder tasked |
| 7 | Logistics management at the staging area is disrupted. | High | • Responder tasked |
| 8 | Higher-than-expected traffic in the area slows all travel and transportation. | High | • Staff travel to staging location<br>• Responder travels to victim with supplies |
| 9 | Higher-than-expected breakdowns of response vehicles occur during operations. | Low | • Responder travels to victim with supplies |
| 10 | Treatment of individuals takes longer than expected because responder training on process had not been done recently. | Medium | • Responder treats victim |

of logistics management disruption affects only one. Reflecting the fact that incident management touches many other functions within the system—through directing their implementation—the medium probability failure mode associated with key leadership being absent affects five other functions.

Though a simple count of how many functions a single failure can hit is a useful piece of information, understanding a failure's real importance requires some idea of the seriousness of its consequences. If a very likely failure has limited impact on performance, it may be worth ignoring no matter how many pieces of the response it will affect. On the other hand, if such a failure has major effects on performance, it might represent the most important target for future preparedness improvement activity.

As a result, assessing how individual failure modes reduce the reliability of response systems requires making an assessment of the type of effects they can have and the level of severity of those effects. Discussion in the previous section introduced the two general effect types—response termination and reduction in capability[28]—as well as the observation that *when* a failure mode occurs during a response is important. Though one could think about a wide variety of different timing characteristics for different failure types, we have simplified down to only two: failures that can occur only at the very beginning of response (which we have labeled *initiation failures*) and failures that can occur at any time during a response operation (labeled *random failures*).[29]

In considering the severity of failure, response-termination failures are easy—they halt all response operations and therefore represent a very high impact from a preparedness assessment standpoint (e.g., the "catastrophic" or "critical" levels of failure severity on the standard FMECA scale quoted earlier in this chapter).[30] For failure modes that reduce response capability, the question is *how much* capability they reduce. Again, any such assessment will likely involve judgment calls and approximation, to

---

[28] Response-capability reductions could arise from failures that

- reduce the ability of the existing number of response assets to operate at their highest efficiency (in this scenario, a reduction in the number of victims that one responder can visit) because of overall delay of response or delay of individual responders acting
- make the actions of the existing of response assets less effective when delivered
- when they occur, reduce the pool of response resources available from that point onward.

[29] Other potential timing profiles include failures that occur early in the operation of a system (so-called burn-in or infant mortality failures) or those that occur late in its operation (wear-out failures.) Though in reliability analysis these failures represent different populations of devices that are failing for different reasons, there are failure modes for response systems that could exhibit these sorts of timing characteristics as well. The easiest example would be failures associated with responder fatigue, which would presumably be more likely to occur late in response operations rather than early in the absence of mitigation such measures as staff rest, rehabilitation, or cycling (see Ebeling, 1997, p. 109 for a discussion).

[30] From the perspective of assessing a response operation intended to save lives and property, the only higher consequence could be a failure that would result in *more* casualties or damage than would have occurred in the absence of response activities. Such an event is possible—e.g., in the later discussion of response to a chlorine response, a botched evacuation that led to a greater fraction of the population being outside and exposed to the hazard than would have been injured in the absence of the evacuation.

provide either a qualitative or an estimated quantitative measure of the impact of a particular failure on response performance.

For the purposes of this example, we will again define a scale similar to (but somewhat more simplified than) the FMECA scale for ranking the severity of failure modes. The categories of the scale are Serious (for failure modes viewed as potentially reducing response capability by 10 to 20 percent), Intermediate (for 5 to 10 percent reduction), Minor (for a 1 to 5 percent reduction), and Negligible (for a less than 1 percent reduction).

Table 2.4 builds on Table 2.3, adding columns for failure mode effect type (response termination or capability reduction), timing, and severity. As was the case previously, even this "intermediate result" of FMECA analysis can inform judgments about the value of different possible preparedness improvements. Failure mode 2 sticks out not only because of its estimated high probability, but also because it is viewed as having a serious impact on performance. Though most of the five consequences for absence of a key member of the response leadership are scored as minor, one is serious—meaning this failure mode warrants serious attention because of both the number of functions it affects and the cumulative effect of its occurrence. Looking at failure mode 4, it is viewed as highly likely that some of the responders included in the plan will not be available, and this is considered an intermediate threat to reliable performance.[31] It is also notable that our example only includes two response-termination failures (which one would reasonably assume to be comparatively rare), but 15 that could reduce capability—largely by creating delays that "eat into" the time available to respond.

### Exploring Quantitative Representations of Response System Reliability

Table 2.4, though simplified and somewhat more qualitative, represents the standard end-state output of techniques such as FMECA. As a summary of failure modes that could affect response performance, with estimates of their probability and effects, such a table can provide one type of snapshot of preparedness concerns. But this type of presentation does not address the third question we posed in the introduction: *How does the importance of individual things going wrong vary for incidents of different size, scale, or complexity?* It is intuitive that every failure mode on the list above will affect the probability of the system being able to serve the full 1,500 people included in its planned $RC_{max}$ (since even a breakdown that cuts response capacity by one person will have an effect when the system is stressed to maximum performance). But for smaller incidents, many failure modes—particularly those with minor effects—will not matter, because

[31] It is easy to see how this particular failure mode could be expanded upon in a more complex analysis. For example, perhaps there is near certainty that 1–5 percent of responders could not be contacted, high likelihood that 5–10 percent would be out of touch, medium likelihood that 10–15 percent could not be contacted, low likelihood for 15–25 percent, and a remote chance for 25–35 percent.

**Table 2.4**
**Qualitative Assessment of Failure Mode Effect, Timing, and Severity**

| Number | Failure Mode | Probability of Occurrence | Functions Affected | Effect | Timing | Severity |
|---|---|---|---|---|---|---|
| 1 | Response communications systems suffer intermittent breakdowns. | Low | Staff dispatched to staging location | Capability reduction (delay) | Initiation | Intermediate |
| | | | Responder tasked | Capability reduction (delay) | Random | Intermediate |
| | | | Responder travels to victim with supplies | Capability reduction (delay) | Random | Minor |
| 2 | Calls from members of the public not needing assistance ("worried well") overwhelm systems. | High | Call for assistance received | Capability reduction (delay) | Random | Serious |
| 3 | A key member of the response leadership is traveling, disrupting the functioning of incident management. | Medium | Notification of incident received | Capability reduction (delay) | Initiation | Minor |
| | | | Staff dispatched to staging location | Capability reduction (delay) | Initiation | Minor |
| | | | Incident management established | Capability reduction (delay) | Initiation | Serious |
| | | | Need assessed | Capability reduction (delay) | Random | Minor |
| | | | Responder tasked | Capability reduction (delay) | Random | Minor |
| 4 | Some responders needed to implement the plan are unavailable. | High | Responder treats victim | Capability reduction (responder numbers) | Random | Intermediate |
| 5 | Supplies that were assumed to be at staging area had been used and not replaced. | Remote | Responder tasked | Response termination | Initiation | |
| 6 | Members of the public cause physical disruption at the staging are, | Low | Responder tasked | Response termination | Random | |
| 7 | Logistics management at the staging area is disrupted. | High | Responder tasked | Capability reduction (delay) | Random | Intermediate |
| 8 | Higher-than-expected traffic in the area slows all travel and transportation. | High | Staff travel to staging location | Capability reduction (delay) | Initiation | Intermediate |
| | | | Responder travels to victim with supplies | Capability reduction (delay) | Random | Minor |
| 9 | Higher-than-expected breakdowns of response vehicles occur during operations. | Low | Responder travels to victim with supplies | Capability reduction (equipment) | Random | Minor |
| 10 | Treatment of individuals takes longer than expected because responder training on process had not been done recently. | Medium | Responder treats victim | Capability reduction (delay) | Initiation | Intermediate |

a system designed for an $RC_{max}$ of 1,500 will have plenty of capacity to absorb the failure and still meet the needs of all victims.

As will become clear in later discussion, we believe that presenting how a response system's reliability varies with the response requirements of different incidents is critical to informing real policy decisions. Whatever the maximum capacity that was chosen to anchor preparedness planning, policy decisions to try to strengthen performance have to consider likely performance across a range of potential incidents rather than simply at the upper end of the spectrum. To address this need, we developed a graphical presentation of response reliability[32] as a function of required response performance (which for our example is the number of victims requiring treatment), from zero (for which the system is 100 percent reliable, by definition) to the $RC_{max}$ (above which system reliability should be zero or near zero, depending on whether performance is viewed deterministically or probabilistically). A system's overall reliability characteristics are then described by the curve connecting those two extreme points.

Figure 2.5 presents three such notional curves. The black dotted line shows a system of perfect reliability; the red line shows a "brittle system" whose performance drops off almost immediately and has essentially no chance of performing anywhere near the $RC_{max}$; and the blue line shows an intermediate system that performs very

**Figure 2.5**
**Illustrative Reliability Curves for Response Systems of Varied Performance**

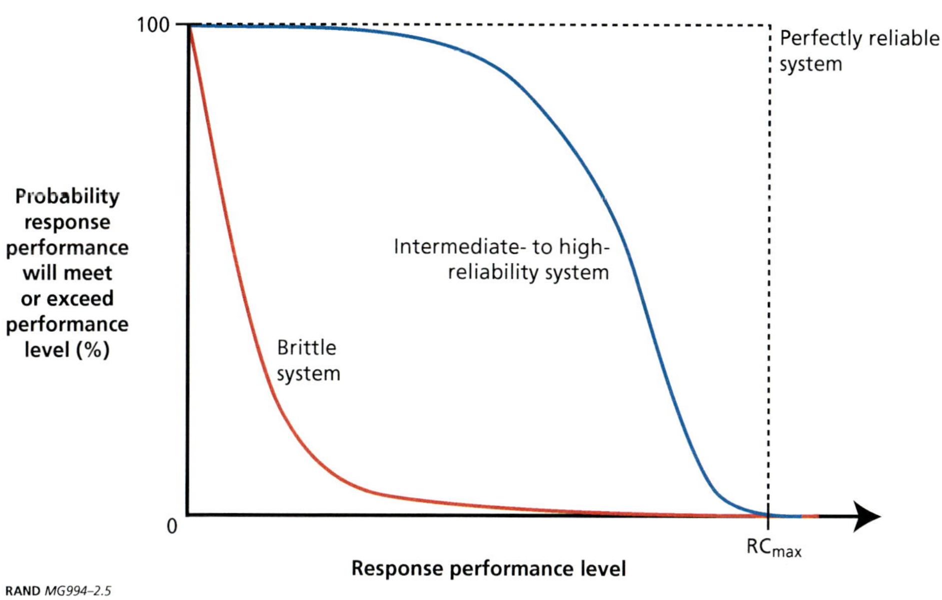


RAND *MG994-2.5*

[32] Again, defined as the probability that the response system (a set of plans, resources, authorities, agencies, and their associated human resources) will be able to deliver at or above a given level of capability at an actual emergency incident.

well up to approximately $1/2^{*}RC_{max}$, after which the probability of it performing well begins falling off.

To actually represent the results of a reliability analysis on a graph like this requires making the transition from qualitative to at least quantitative *estimates* of the probability of failures occurring and their effect on response performance. For example, for this simple example system, in which response performance is essentially the production of an output (treatments) over time, translating a failure mode into a probability of losing the capability to deliver a specific number of treatments makes it more straightforward to determine the incidents at which that loss will be important versus those at which it might not be important at all. The different types of failure modes that we presented earlier—response-termination failures and capability-reduction failures occurring either at the beginning of response (initiation failures) or randomly during the operation—each have a different effect on response reliability as represented on such a curve. Before we show what such a curve would look like for our example response system, with all of the ten failure types discussed above, we will illustrate, one at a time, how different failures types affect the shape of a response reliability curve.

To build these curves, we used a basic statistical model run in Microsoft Excel that simulated response performance as a system that delivered treatments over time at an average rate. We generated actual performance in a simulation run using a Poisson distribution around that average rate, chosen to result in a system that would deliver 1,500 or more treatments at its 90th percentile of performance.[33] We then generated the different failure types using the internal random number generator in Excel, based on inputted probabilities of failure occurrence. For failures that could occur at any time during the response operation, their time of occurrence was similarly randomly generated. We represented the effects of all failures as a reduction in treatment rate (by some percentage for capability-reduction failures, or *in toto* for response-termination failures).

As described earlier in this chapter in our initial description of reliability engineering and analysis, this approach combines the two central elements of such analysis: an "engineering-modeling" component based on an understanding of the system and how it functions (the identified failure modes and their estimated effects) and a probabilistic element (modeling variation in response performance using a random distribution around a mean), acknowledging that we do not believe we are (or even could) capture all possible sources of variation in performance as explicit failure modes.

For each analysis of the effect of one or more failure modes, we generated 1,000 simulated response operations. For those in which failures occurred, the initial treatment rate might be reduced and/or the total response time might be divided into one

[33] This involved an average treatment rate of 14.5 patients per time step in a 100-time-step response operation. The use of random variation in response performance means that, in the simulations, the total treated could exceed 1,500 for some cases. A histogram of the response performance in 1,000 simulation cases of the response system with no failure modes is included in each of the subsequent figures for comparison to cases with different types of failure modes.

or more periods of different treatment rates. We simulated the response output for each period separately, using another random value that defined where on the Poisson distribution the performance level would be chosen. We then calculated the response output of the case by totaling the number of treatments produced for the entire response. We built output histograms for the cases and calculated exceedence curves describing the probability of performance exceeding each output level from 1 to 1,600 treatments.

The following subsections demonstrate the effects on the response reliability curve of each class of failure mode.

**Response-Termination Failure at Initiation.** The easiest situation to illustrate is the effect of a failure that could occur at the beginning of the response that will halt response operations entirely. This type of failure would make it impossible to produce any response output at all, and so affects the ability of the response system to respond to incidents of any size. Since such a failure would uniformly cause the response to fail, it would act to "push down" the response reliability curve for the response system by the full probability of its occurrence (in this example, by 10 percent) for all incident scales.

The effect on our example response system is shown in Figure 2.6. In the simulation results, the histogram of the response operation with the failure mode differs from the base case in the appearance of cases with no response output, and there is a uniform reduction in reliability across all incident sizes.

**Random Response-Termination Failure.** Moving away from failures at the beginning of the response operation brings in the added complexity that even serious failures occurring late enough (after the response is well under way) will have a much-reduced effect on overall performance. To illustrate their effect, we will first discuss response-termination failures that might occur at some later point in a response operation. Unlike the initiation termination failure, whose effects were the same irrespective of incident size, there is a clear size effect here: The effect on response reliability is greatest for large incidents since, close to the $RC_{max}$, even a very late termination would reduce the chance of the response meeting a high performance threshold. At smaller incidents, the effect on reliability will be less because of the possibility that the response-termination failure could occur late enough that the needs of the response will already be met.[34]

Figure 2.7 shows the effects of such a failure on response reliability. In this case, unlike the cluster of cases at zero performance seen for initiation failures, the cases in which the failure occurred are spread over the entire histogram. This produces an

[34] Since each response case in our simulation has a constant duration, we are calculating response output over the entire time for every case—and when we consider the probability of a failure occurring, it is the probability of its occurrence over that entire time period. In reality, a small incident would be resolved and its response completed in only a portion of the time period we are simulating and so there would be a shorter time over which failures could occur (assuming they were random over the entire time). For the purpose of our reliability calculations, this is essentially equivalent to our saying that a failure might occur after the needs of a small response have already been met.

**Figure 2.6**
**Effect of an Initiation Response-Termination Failure Mode on a Response Reliability Curve**

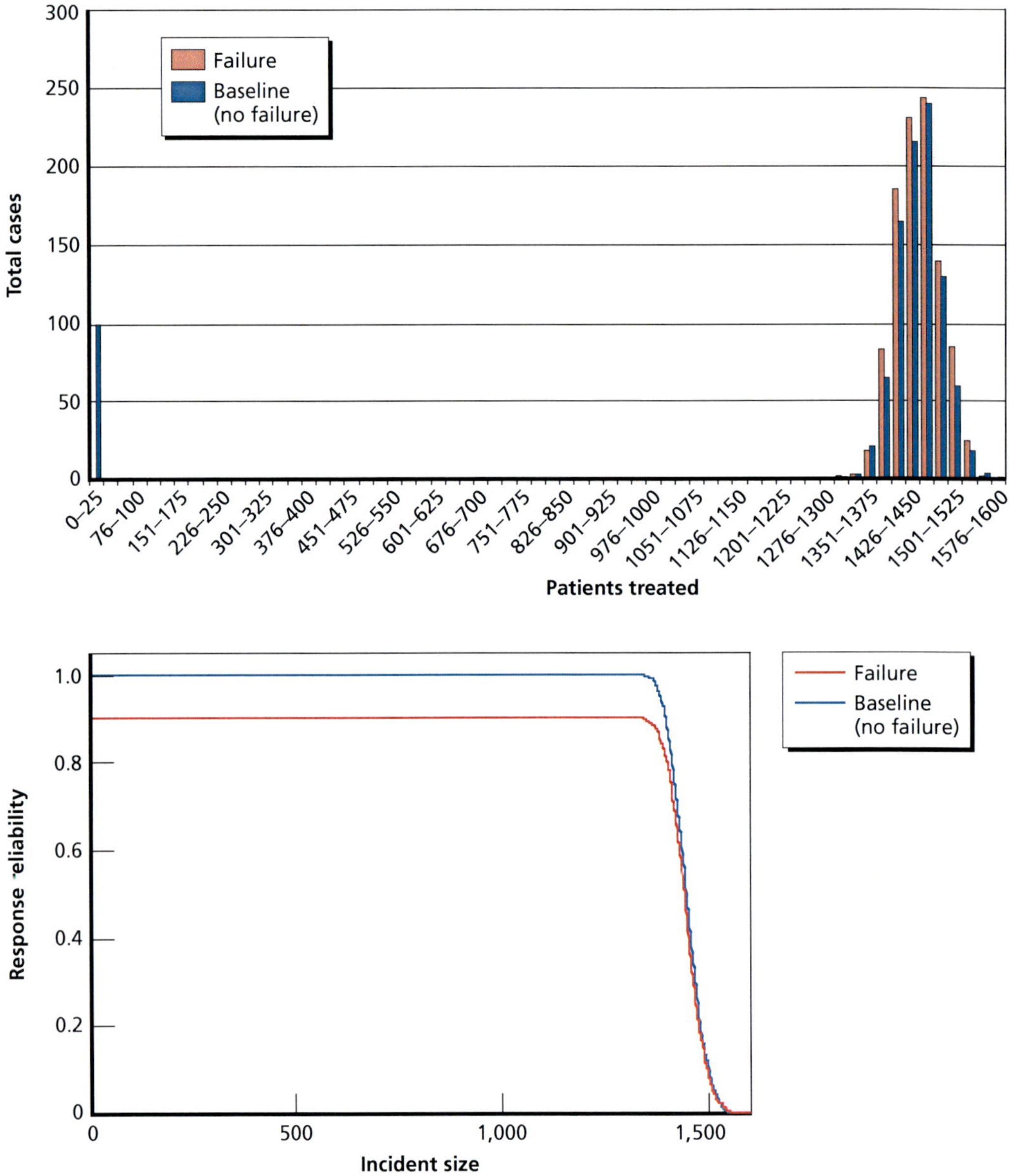


NOTE: Average treatment rate of 14.5 per time step for base case; one initial response-termination failure with a 10 percent probability of incidence that reduced treatment rate to zero.
RAND *MG994-2.6*

**Figure 2.7**
**Effect of a Random Response-Termination Failure Mode on a Response Reliability Curve**

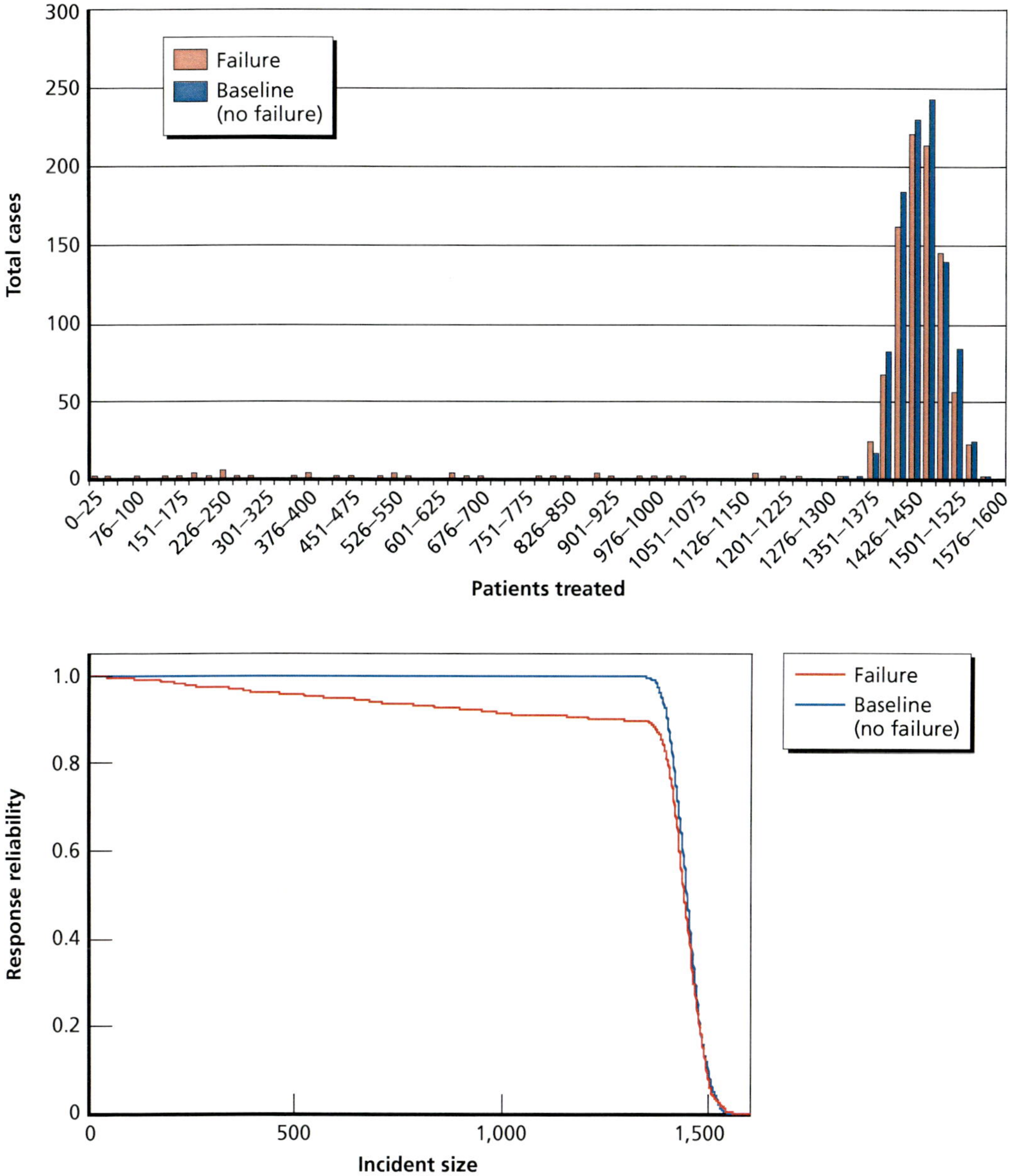


NOTE: Average treatment rate of 14.5 per time step for base case; one random response-termination failure with a 10 percent probability of incidence that reduced treatment rate to zero.

RAND *MG994-2.7*

effect on response reliability where performance is reduced by the full probability of its occurrence (in this example, 10 percent) at the right of the graph, but drops essentially linearly (given the random distribution of failure times) to zero for very small incidents.

**Capability-Reduction Failure at Initiation.** Though some failures that could occur at the beginning of the response would affect the ability to respond at all, others will just have a capability price associated with them—e.g., because of a shortfall in training the response is slightly less effective than it would have been otherwise.

Unlike the initiation response-termination failure, which affected incidents of all sizes equally, the effects of a reduction in capability at the beginning of the response on the reliability curve depend on how large that reduction could be. To illustrate the effects, we simulated the effect on system performance of an initiation capability-reduction failure that had a 10 percent chance of occurrence, and if it did occur reduced the system output by approximately 10 percent (Figure 2.8). For failure modes that occur at the beginning of response but affect only total system capacity, effects on reliability are seen only on the far right of the graph above. For the 10 percent of the responses affected by the failure, a separate population of responses at reduced output is produced (the "shoulder" observed in the histogram).

**Random Capability-Reduction Failure.** Capability-reduction failures could also occur randomly during a response operation. Like random response-termination failures, the effect of these sorts of failures on response reliability for smaller-scale incidents will be attenuated the later they occur in the response operation. As a result, they will have a greater effect on the reliability graph moving from right to left (i.e., increasing incident size). To illustrate the effect of this failure mode, we used the same case as above for the initial capability-reduction failure (10 percent chance of occurrence, just over a 10 percent capability reduction if it did occur), but for the situation in which it could occur at any time during the response. Figure 2.9 shows the result. When the capability-reduction failure can occur at any time in the response rather than just at the beginning, its effect on response output—and therefore reliability—is more attenuated. Rather than suffering a reduction in rate for the whole operation, failures that occur late could reduce total performance very little. As a result, the distortion in the histogram, and of the reliability curve, is more modest than the previous case.

**Multiple Failure Mode Effects on Response Reliability.** Similar simulations to those illustrated above can be run modeling the effects that multiple possible failure modes can have on the reliability of a response system for different size incidents. Addressing the effect of multiple failure modes on a response system's reliability is somewhat complex, given the possibility of multiple failure modes occurring in a single response (e.g., an initial capability-reduction failure followed by later random failures that either reduce response effectiveness or terminate operations).[35]

---

[35] We discuss an approximate way of producing these curves without probability modeling in Appendix A of this document.

**Figure 2.8**
**Effect of an Initiation Capability-Reduction Failure Mode on a Response Reliability Curve**

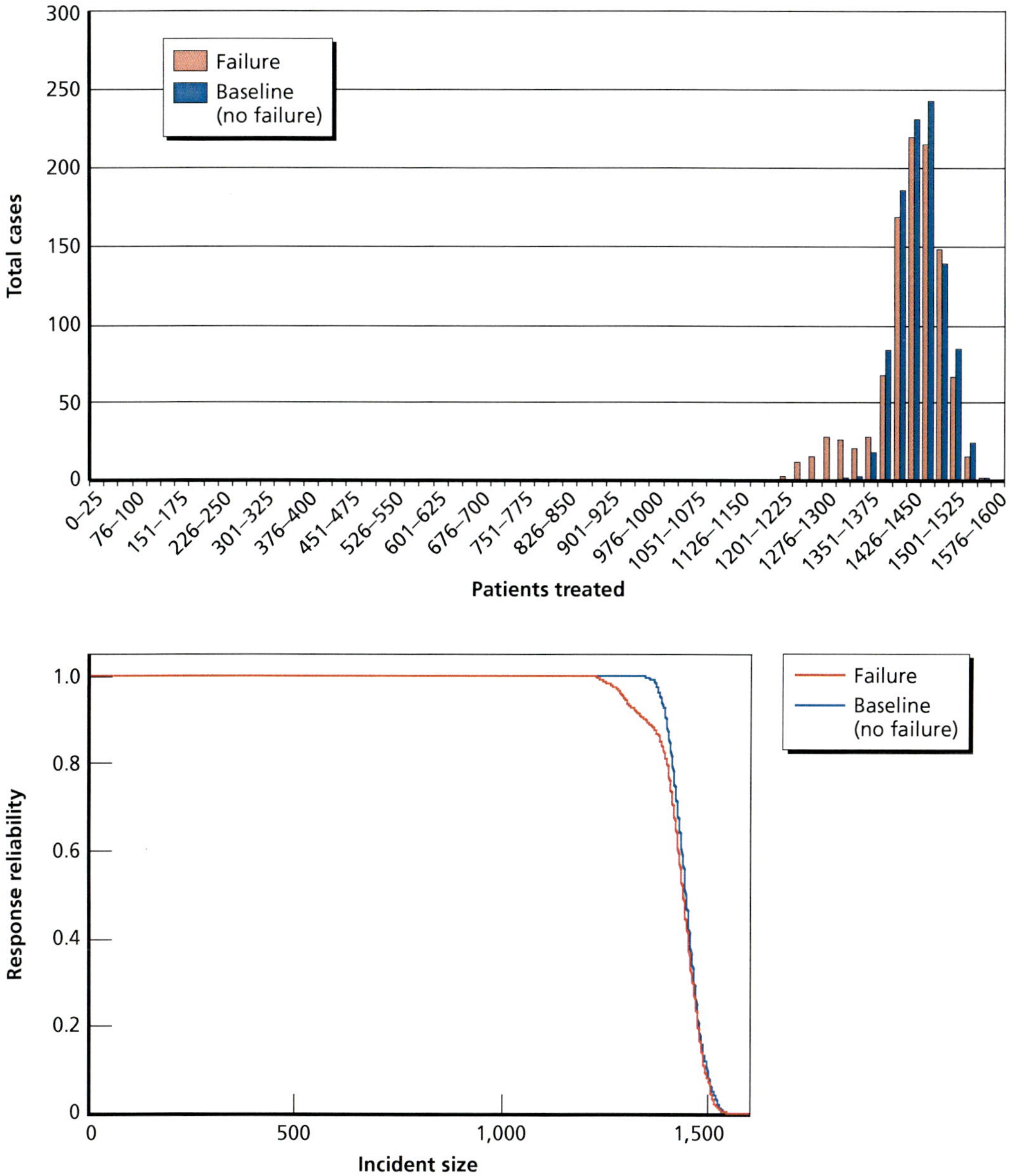


NOTE: Average treatment rate of 14.5 per time step for base case; one initial capability-reduction failure with a 10 percent probability of incidence that reduced treatment rate by 1.5 when it occurred.
RAND *MG994-2.8*

**Figure 2.9**
**Effect of a Random Capability-Reduction Failure Mode on a Response Reliability Curve**

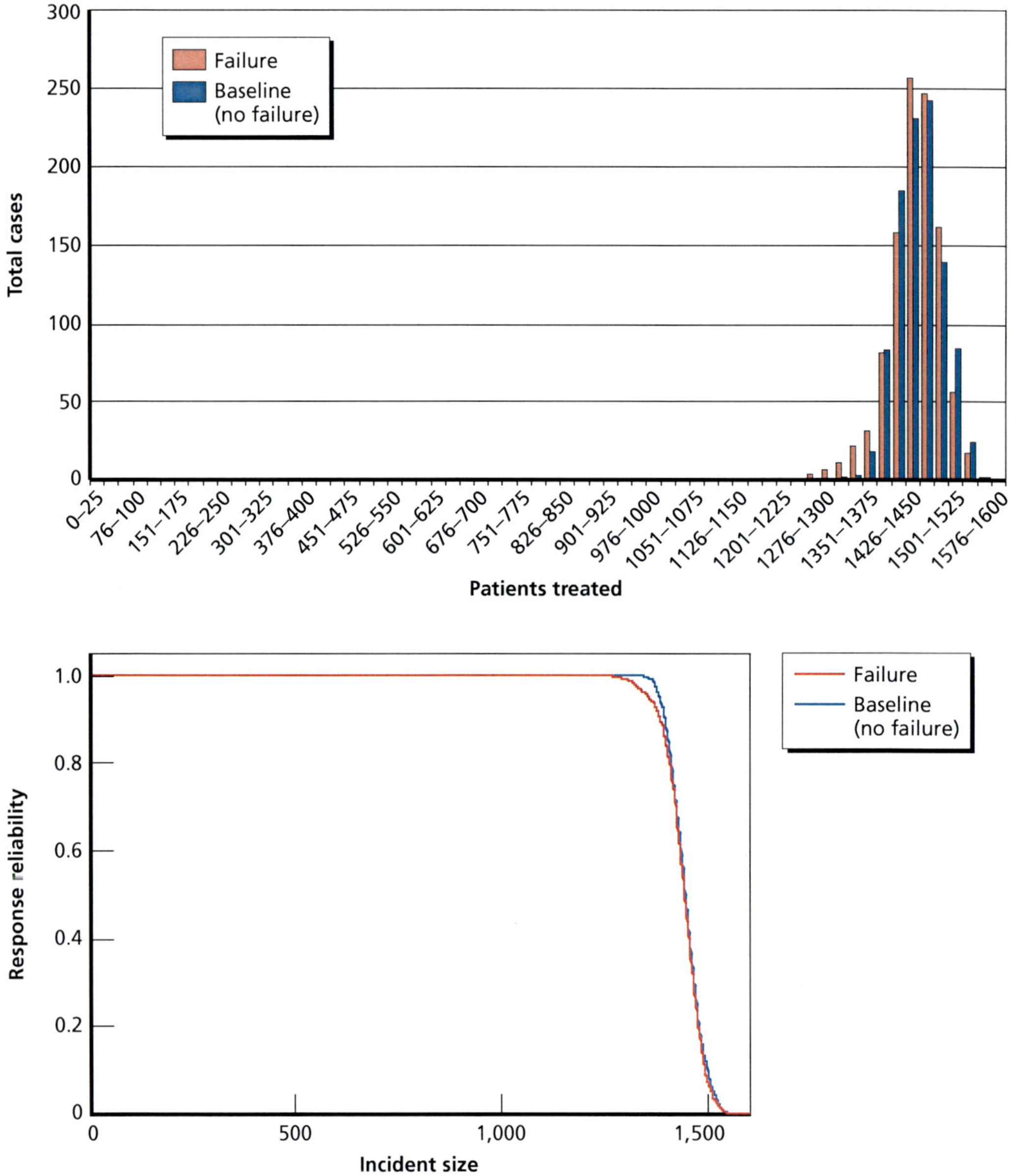


NOTE: Average treatment rate of 14.5 per time step for base case; one random capability-reduction failure with a 10 percent probability of incidence that reduced treatment rate by 1.5 when it occurred.
RAND *MG994-2.9*

To illustrate the results of such an analysis, we will return to our example response system, with its ten failure modes discussed previously. To construct a curve for that system's performance, we need to convert the qualitative measures of probability and consequence discussed earlier into quantitative representations of those measures. Table 2.5 reproduces the results in Table 2.4, but replaces the qualitative rankings with the midpoint of the range associated with the ranking in the text (numbers shown in shaded cells). Though we can simply make this substitution for our illustrative example, going from qualitative to quantitative estimates in a real system would require the use of other methods to obtain better estimates, or—if approximate numbers were used—sensitivity analysis to assess how changes in the numbers would affect any conclusions drawn based on the results.

The overall curve for the reliability of this response system is shaped by the probabilities and consequences for all the modes in the table. For failure modes that affected different parts of the model, a single occurrence of the failure produced all of the later effects (as discussed above). To make it more obvious how the different failure modes "build up" to a composite performance picture for the system, we ran independent simulations as failure modes were sequentially added to the system. The order in which they were added was determined by their type, with termination failures added before capability-reduction failures, and with failures that affected many steps added last. Figure 2.10 shows the resulting progression of graphs as failures were added in the following sequence, where the numbers correspond to those included in Table 2.5: (5), (6), (7), (10), (2), (4), (9), (3), (1), and (8).

Such a curve for our example system presents a composite snapshot of its likely performance across different incident sizes. With its ten failure modes of varying probability and generally modest impact on performance, the greatest effect on response reliability is seen for larger incidents, starting approximately at 1,000 victims (or two-thirds of the system's $RC_{max}$). Below that level, though affected by response-termination failures and the possibility of multiple simultaneous failures reducing performance, the reliability of the system approaches 100 percent.

## Response Reliability Measures Applied to Preparedness Policy Problems

Though identifying and analyzing failure modes in a response system can make very tangible contributions to improving emergency preparedness planning, we also believe that this approach—particularly when it can be done quantitatively—has the potential to directly inform a number of other, much broader policy questions relating to emergency preparedness. One of our rationales for walking through an example analysis in detail was to set up this broader discussion, making it possible to demonstrate the value of the approach if it can be effectively and practically performed on real

**Table 2.5**
**Notional Quantitative Estimates for Failure Probabilities and Consequences**

| Number | Failure Mode | Probability of Occurrence (%) | Functions Affected | Effect | Timing | Severity (%) |
|---|---|---|---|---|---|---|
| 1 | Response communications systems suffer intermittent breakdowns. | 1.5 | Staff dispatched to staging location | Capability reduction (delay) | Initiation | –7.5 |
| | | | Responder tasked | Capability reduction (delay) | Random | –7.5 |
| | | | Responder travels to victim with supplies | Capability reduction (delay) | Random | –3 |
| 2 | Calls from members of the public not needing assistance ("worried well") overwhelm systems. | 7.5 | Call for assistance received | Capability reduction (delay) | Random | –15 |
| 3 | A key member of the response leadership is traveling, disrupting the functioning of incident management. | 3.5 | Notification of incident received | Capability reduction (delay) | Initiation | –3 |
| | | | Staff dispatched to staging location | Capability reduction (delay) | Initiation | –3 |
| | | | Incident management established | Capability reduction (delay) | Initiation | –15 |
| | | | Need assessed | Capability reduction (delay) | Random | –3 |
| | | | Responder tasked | Capability reduction (delay) | Random | –3 |
| 4 | Some responders needed to implement the plan are unavailable. | 7.5 | Responder treats victim | Capability reduction (responder numbers) | Random | –7.5 |
| 5 | Supplies that were assumed to be at staging area had been used and not replaced. | 0.5 | Responder tasked | Response termination | Initiation | –100 |
| 6 | Members of the public cause physical disruption at the staging are, | 1.5 | Responder tasked | Response termination | Random | –100 |
| 7 | Logistics management at the staging area is disrupted. | 7.5 | Responder tasked | Capability reduction (delay) | Random | –7.5 |
| 8 | Higher-than-expected traffic in the area slows all travel and transportation. | 7.5 | Staff travel to staging location | Capability reduction (delay) | Initiation | –7.5 |
| | | | Responder travels to victim with supplies | Capability reduction (delay) | Random | –3 |
| 9 | Higher-than-expected breakdowns of response vehicles occur during operations. | 1.5 | Responder travels to victim with supplies | Capability reduction (equipment) | Random | –3 |
| 10 | Treatment of individuals takes longer than expected because responder training on process had not been done recently. | 3.5 | Responder treats victim | Capability reduction (delay) | Initiation | –7.5 |

**Figure 2.10**
**Composite Response Reliability Curve for Our Example Response System**

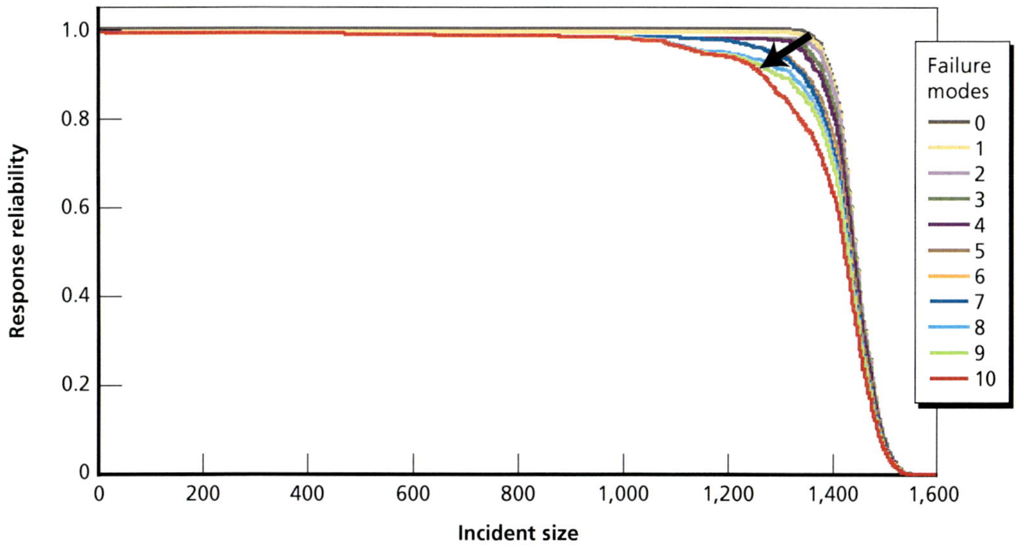


NOTES: Each line represents the results of 1,000 response simulations. Starting conditions for each simulation were identical, with the exception of the sequential addition of failure modes (of type, probability, and consequence, as described in Table 2.5), as described in the text.
RAND *MG994-2.10*

emergency response systems and preparedness plans (a topic we will turn to in the remaining chapters of this document).[36] In considering how this analysis could inform deliberation in several different policy areas, we will build from questions for which the qualitative, tabular results of the FMECA analysis alone are useful up to questions for which the response reliability graphs and the (pseudo)quantitative comparisons they can enable could be of particular value.

In the remainder of this chapter, we will discuss how measures of response reliability can contribute to

1. qualitatively prioritizing among possible preparedness investments
2. informing trade-offs between actions that would improve performance for large incidents versus smaller, more common events

---

[36] In considering building from this simple example to analyses of real response systems, the reader should keep in mind that most real response activities will have multiple outputs—not just the single measure discussed here. In our later discussion of a chlorine response operation, for example, at the minimum two such reliability graphs would be needed to describe the outputs of response activities. Demands for resources to ensure reliable performance for one capability might compete with other response actions requiring trade-offs. Such competition would essentially be an additional failure mode that all response functions would have to deal with, compared with the simple case we have described in this chapter.

3. enabling quantitative prioritization and assessment of the cost-effectiveness of different preparedness improvement options
4. answering the question "How much is enough?" with respect to preparedness investments.

### Prioritizing Possible Preparedness Investments

In an engineering context, the output of FMECA analysis is used to identify and prioritize reliability problems so they can be fixed during the design stages of system's development. As discussed above, the "list output" of failure modes and qualitative assessments of their probability and consequence can serve a similar function for identifying and prioritizing among possible changes in preparedness planning. All other things equal, failure modes with higher probabilities, affecting performance in more parts of the response system, and with higher consequences would be higher priorities for corrective action.

In this way, the *process* of the reliability analysis—mapping response activities as a system, cataloging what might go wrong, identifying how failures would affect one or more components of the system—becomes an adjunct to prudent planning processes and red teaming that might be done in the creation or revision of a preparedness plan. In doing so, the FMECA process of expressing the consequences of different failures in terms of their likely consequence for system performance (rather than some intermediate impact on system functioning that could differ considerably from failure mode to failure mode) provides a more common basis for comparison.

### Making Trade-Offs Between Actions to Improve Performance for Large-Scale Incidents Versus Smaller-Scale, More Common Events

Planning to mitigate failures in emergency response systems generally requires setting priorities and making choices. Addressing every potential failure mode would be one possible strategy, but it would be an expensive one. As discussed previously with respect to engineering design and failure mode analysis, there is nearly always a trade-off between improving reliability and increasing cost.

Furthermore, fixing *every* potential failure mode in a response system would be beneficial in responses to the largest-scale and most demanding incidents. For example, considering the data presented in the illustrative response reliability graph (Figure 2.10) for our example system, fixing all of the ten failure modes would have very little benefit *except for the incidents to the far right of the graph.* For the illustrative system with all ten of its problems, reliability is already very high up to incidents with 1,000 patients, and is above 90 percent even at 1,200 to 1,300 patients. The major impact of the failure modes we posited for that system is for very large-scale incidents, for which reliability falls sharply. This example is consistent with the observation that most response systems perform very well for most incidents—and that major problems arise only after an incident passes a size threshold (Miskel, 2008).

Should a policymaker considering our example system allocate resources to address any or all of its ten identified failure modes? There is no single correct answer to that question, but assessing that system's reliability as incident size increases (and, therefore, as required response performance increases) shows clearly that additional investments should be made only if the goal is improving large-scale incident performance. If doing so is a political or policy priority, the list of failure modes (and judgments about which are more likely or consequential) provides guidance on where to start. But if the main concern is reliable performance for smaller-scale incidents, or if the likelihood of a very large-scale incident occurring is viewed as sufficiently low that it does not warrant additional attention, the answer may be that no further investments need to be made. Analyzing and assessing reliability in this way makes that choice clear, and could contribute to explaining the reasons behind—and desired outcome of—any resource allocation that was subsequently made.

### Comparing the Cost-Effectiveness of Different Preparedness Improvement Options

Beyond showing the effect of different failure modes on performance at different incident sizes, the information in these response reliability curves—specifically the area under a system's reliability curve (e.g., Figure 2.10)—can be used as an aggregate measure of the performance of the system.[37] Such a measure can be applied in three main ways to answer different questions about preparedness performance or approaches to strengthen preparedness.

*First, it can be used to compare the performance of a system with a set of identified failure modes against a theoretical system without them* (illustrated in Figure 2.11).[38] This can provide a way to assess—however many failure modes have been identified—their total effect on system performance. In the case of our example system, the value at which there are no failure modes (as one might expect) approaches 1,450 treatments

[37] The use of this value as a measure of system performance depends on the validity of the estimates of the probabilities and consequence values underlying the curves themselves. In an ideal situation, quantitative estimates could be made for every failure mode, allowing creation of a precise and accurate reliability curve for the system being assessed. Data limits (as we will discuss in later chapters) could mean that approximations would have to be made that would make the specific value of the measure less reliable as a description of the system itself. However, to the extent that estimates across failure modes could be made *consistently*, even approximate measures might still be useful when comparing different policy options for addressing failure modes and weighing the *relative* (if not absolute) cost-benefit of different courses of action.

[38] To calculate that measure for our example system, we have used the fact that our simulations return a probability of performance exceeding a given number of treatments that goes from zero to the maximum capacity of the system increasing by one treatment at a time. As a result, we have approximated the area under the curve by simply adding up the y-values for the entire graph (essentially treating the area as made up of rectangular slices that are each one "treatment" wide). In practice, the possibility of random variation producing more than 1,500 treatments for this system means that we performed this rough calculation from zero (where all systems perform with 100 percent reliability) up to 1,600 to ensure that we captured the performance of the full system.

**Figure 2.11**
**Area as a Relative Measure of System Performance**

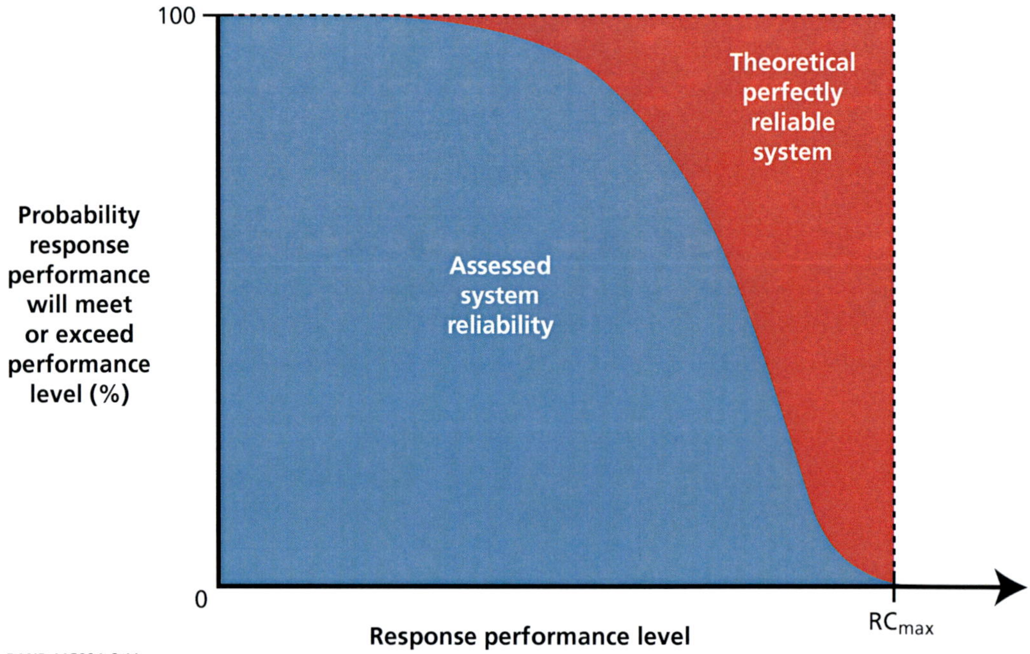


RAND *MG994-2.11*

delivered given the average treatment rate and the length of the response window.[39] With all ten of the failure modes (Table 2.5) included in the simulated system, total performance drops to 1,382 expected treatments, a relatively modest reduction of between 4 and 5 percent.[40]

*Second, such a quantitative measure is good starting place for considering the different improvement options available if policymakers want to strengthen preparedness.* How much a policy or preparedness measure would increase that area—e.g., an intervention to fix one of the system's failure modes—provides a measure of its value.[41] Starting from our sample simulation including all ten failure modes, we chose several modes and posited

---

[39] Variation in response rate as a result of the random elements of the model resulted in an actual simulated value of 1,448 treatments.

[40] This modest drop in aggregate performance reflects the relatively modest scale of the failure modes we included in the example. If we multiply all the failure modes' probability and impact by 5 (except the response-termination failures, whose consequence remained the same)—representing a response system with much more serious performance problems than our example case—the analogous value is 549, a drop of more than 60 percent from the theoretical maximum performance.

[41] Changes in policy or preparedness efforts obviously could also *reduce* the area under the curve if they increase the probability or consequences of existing failure modes, or introduce new ones. The logic in this section applies equally to examining such changes (whether intentional or not), though we have framed our discussion around comparing changes intended to increase preparedness.

policy interventions that "fixed" them one at a time. In our simulation, the practical implementation of this was simply zeroing out the probability that those failures would occur one at a time, doing several simulations of system performance with different sets of nine failure modes. We identified a subset of the failure modes that included both response-termination and capability-reduction failures, and ones that affected multiple as well as single parts of the system. In each case, a simulation was done as described previously and the area under the resulting reliability curve calculated and compared to the performance in the "base case" including all ten failure modes. The difference between the resulting values and the base case, both in the number of units as well as in the percentage of performance improvement, is shown in Table 2.6.

The units (corresponding to an increase in the area under the curve by 1) reflect the aggregate effect of the probabilities of the failures being addressed, the amount of their consequences, and the number of parts of the system affected by their occurrence. As a result, such values could be used to score the relative value of different options relative to one another. But looking just at improvements in reliability can be deceptive in some ways—particularly for a system like our example case, in which performance even with all failure modes included is relatively robust. As a result, the next column calculates what percentage improvement over the base case those additional units represent—values ranging from under 0.5 percent to almost 1.5 percent.

Though fixing the failure modes that hurt reliability is one strategy for improving the performance of this system for large-scale incidents, it is not the only strategy. An alternative approach to strengthening preparedness could be to simply buy more system capacity—in our simple example, by hiring more staff and associated material required to be able to treat more people more quickly. This intervention would essentially shift the entire curve for the system outward, by moving the distribution of performance up and, therefore, improving reliability for a subset of incident sizes. In Table 2.6, we show an exemplary case in which we increased the average treatment rate in our example system from 14.5 to 14.75—a just under 2 percent increase in total system capacity—and the calculated effect on total system reliability. In this case, simply adding a small amount of additional response resources produced a reliability increase of more than 1.5 times the size of the best option we looked at that involved fixing reliability problems in the system.[42] Such an increase would also result in the possibility of responding to larger incidents (since such an increase pushes up the $RC_{max}$ of the system) in addition to the increase in reliability for smaller responses. As a result, this measure can provide a common basis for comparing very different preparedness interventions.

---

[42] This conclusion is dependent on the characteristics of our example system and its reliability problems. Above we cited a simulation where we had increased both the probability of occurrence and the consequences of our failure modes (except response-termination failures since their consequence was already a 100 percent reduction in performance). Its baseline performance was only 549. In that simulation we then compared the benefits of fixing failure mode 2 (policy option 2 in Table 2.6) with increasing the capacity of that system by 0.25 (the last policy option in Table 2.6). In that case, fixing the failure mode dominated, producing almost six times the total reliability improvement compared to the capacity increase.

**Table 2.6**
**Using Response Reliability Values to Compare Preparedness Improvement Options**

| Policy Option | Performance Improvement Over Base Case: Units | Performance Improvement Over Base Case: Improvement (%) | Exemplary Annual Cost ($, thousands) | Cost/ Unit Increase ($, thousands) |
|---|---|---|---|---|
| Eliminate Failure Mode 1: Response communications systems suffer intermittent breakdown | 5 | 0.4 | 500 | 100 |
| Eliminate Failure Mode 2: Calls from members of the public not needing assistance ("worried well") overwhelm systems | 17 | 1.2 | 250 | 15 |
| Eliminate Failure Mode 3: Key member of response leadership traveling, disrupting functioning of incident management | 19 | 1.4 | 150 | 8 |
| Eliminate Failure Mode 6: Physical disruption at staging area caused by members of the public | 9 | 0.7 | 50 | 6 |
| Eliminate Failure Mode 7: Logistics management at staging area disrupted | 6 | 0.4 | 250 | 42 |
| Increase response system capacity by almost 2 percent (increase average treatment rate from 14.5 to 14.75) | 33 | 2.4 | 200 | 6 |

NOTES: Units reported reflect the difference between the greater area under the response reliability curve for the case with the policy change and the base, ten-failure-mode case without it, rounded to the nearest unit. Percentage improvement is reported over the base case performance of 1,382 discussed in the text. The "exemplary annual cost" column is an entirely illustrative number generated for each option only to allow demonstration of use of this measure for comparative cost-effectiveness assessment. Measures are rounded to the nearest $1,000.

*Finally, comparisons of the areas under the reliability curves for different policy options can be a starting point for cost-benefit or, at the minimum, relative cost-effectiveness analysis.* Responsible emergency response planning must involve consideration of costs. In the remainder of the table, we explore how this measure could be used as a component in assessing the comparative cost-effectiveness of different policy interventions. We assigned notional prices to each of the options to illustrate how these values could be used to construct a cost-effectiveness metric, the final column showing the cost per unit of reliability improvement. In an actual comparison of alternative policies, supported by real cost data for addressing different failure modes and alternative capacity-improvement options, this measure could help to identify which options would provide the greatest reliability improvement at least cost—and therefore represent the most attractive targets for the marginal preparedness dollar. In our illustrative example, the option of slightly increasing system capacity, though more expensive than some other choices, is competitive with fixing some failure modes—and since it provides the additional benefit of slightly improving performance for incidents above the $RC_{max}$ of the base case system, it would presumably be the preferred option.

**How Much Preparedness—and Response Reliability—Is Enough?**

But beyond helping to allocate preparedness dollars to the most effective strategies, these measures could also help to frame debates of the broader question of "How much preparedness is enough?" Determining how much preparedness is enough involves, at a minimum, answering three questions. As we have framed it here, the first question is *How large an incident should a jurisdiction, area, or the country overall be prepared to respond to?* Essentially, What is the appropriate value for $RC_{max}$? In considering how much a jurisdiction, area, or the country overall wants to spend on preparedness efforts, defining the desired upper limits of system performance is an important factor and has been part of the focus of such efforts as DHS's development of National Planning Scenarios (DHS, 2005) and the TCL (DHS, 2007b). However, our reliability analysis clearly shows that it is not the only—and perhaps not even the most important—factor in the question of how much preparedness is enough.

The second question that our analysis suggests must be asked in concert with the size of the incidents we are preparing for is *How reliable should response systems be for incidents of various sizes?* As this discussion has shown, the question of making investments to achieve a desired $RC_{max}$ and target reliability characteristics for a system are closely related and can potentially be pursued simultaneously through common policy approaches, but they are still distinct questions. In some cases, it may be good policy to focus investments on increasing response capacity, since doing so will also increase reliability for smaller-scale incidents. But if the central goal is improving the chances the response system will function well at future incidents, more-focused efforts aimed at key failure modes could be a better strategy.

The third and perhaps most important question is *How much should we be willing to pay for capacity or reliability over what we have now?* In our example case, even the most cost-effective options for investment of the marginal preparedness dollar would produce only a few percentage points' worth of improvement, since the performance of the baseline system, even with its ten failure modes, was already quite high.[43] With a cost of \$50,000 per 0.7 percent improvement over the status quo for this system, it would be legitimate to ask whether paying for that improvement is the best use of those resources. As our example here suggests, there will be a point at which the performance

[43] This is separate from the question of whether the $RC_{max}$ of treating approximately 1,500 people that we used in this example is the "right" performance ceiling for this system. The question of how much is enough relates to both the maximum capacity and desired level of reliability of a system with a given $RC_{max}$ value.

of a system is sufficiently high that the resources required to address its residual failure modes will yield only small marginal increases in its performance.[44]

In contrast, a system whose performance is much further below its theoretical ceiling performance could get a much larger absolute and percentage boost from fixing a particular failure mode, potentially resulting in a stronger rationale for additional preparedness investment. A common measure of the reliability improvement achieved by addressing different failure modes within a response system, or by just adding additional resources to a system whose capacity or reliability thought to currently fall short, not only makes it possible to compare options and ask which of them is the best, but also provides a yardstick to compare the range of possible end states with the status quo and ask the broader question about the value of additional investment compared with other possible uses of those resources.

---

[44] This is similar to the argument made in the classic book by Enthoven and Smith regarding the payoff of marginal increases in additional military forces. See Enthoven and Smith, 2005, Chapter Six: Yardsticks of Sufficiency, pp. 206–242).

CHAPTER THREE

# Describing a Chlorine Release Scenario and Relevant Response Parameters

To develop an approach for measuring the reliability of a response operation, it is necessary to examine a realistic emergency scenario requiring more complex response operations than the simple example described in Chapter Two. The characteristics of an incident define what the requirements are for response operations and what "success" at responding would mean. This includes the specific capabilities[1] that are necessary to respond and the scale of those capability requirements. The specifics of a particular emergency or disaster also shape the variety and importance of potential complications that could hinder response efforts and get in the way of responders implementing a response plan "as written."

As a test case for our methodology, we selected as our incident scenario a significant chlorine release from an industrial-size tank. Such a scenario has a number of advantages. As later discussion will show, responding to a chlorine release incident requires a number of different response capabilities over varied timescales, which could potentially be traded off against one another. This scenario also can affect a significant area, but not so broad an area to be too complex for this sort of prototyping analytical effort.

The following sections describe our chlorine release scenario and discuss the implications of the scenario for considering the analysis of response capabilities and requirements.

[1] When we use the term *capabilities* in our text, we are referring to the ability of responders to perform specific tasks in a response operation. We are using this term in a roughly equivalent way as FEMA doctrinal documents, such as the TCL (DHS, 2007b), though in some cases we do not define the "boundaries" of individual capabilities identically with the TCL. In Appendix B, we lay out how the capabilities we use in our analysis correspond to those in the TCL.

## Describing a Chlorine Release Scenario

A chlorine release, whether intentionally created by a terrorist or criminal seeking to cause harm or resulting from an industrial accident or negligence, consists of a hazard originating at an initial site that can then travel with prevailing winds and cause harm to areas away from the source site.[2] Though physical damage will certainly occur at the originating site (e.g., at the minimum, the damage necessary to cause the release), and can also occur in areas exposed to high concentrations of the gas, the primary concern in a chlorine release is the toxic effects of the gas on people or other living things. At high concentrations, chlorine exposure can rapidly be lethal, with lower concentrations potentially producing serious respiratory and other injuries (Table 3.1). Incidents in which gaseous chlorine release does not occur instantaneously, but instead leaks in liquid form and then vaporizes, produce a time course such that an initially high-concentration cloud of the gas forms at the source site and, as it travels, gradually disperses to a point where eventually it will reach levels that are no longer dangerous to life or health.[3]

For the purposes of illustration, Figure 3.1 shows the notional evolution of an extended-release chlorine incident. To ease presentation, the incident is shown as progressing in two dimensions from left to right in the figure. However, after the breach of the chlorine source (pictured here as a generic industrial tank of appropriate characteristics and volume to produce an extended cloud release), the first area affected is the immediate proximity of the source site. Along available dispersal routes for the gas, the cloud will travel with the local airflows, remaining close to ground level. As a result, from the time of release (with individual diagrams moving down the figure showing the advance of time), a set of *threatened sites* are created that may be within reach of the gas dispersal. As the cloud moves, threatened sites are converted into what we have labeled *affected sites* as they are exposed. With time, the gas cloud will eventually dis-

---

[2] Chlorine is generally transported and stored as a liquid either under high pressure or at low temperature. At atmospheric pressures, chlorine's boiling point is –34°C. Pressurized storage and transport is more common (Barrett, 2009). Chlorine is most often contained in 100- or 150-pound cylinders, 1-ton containers, or 55- or 90-ton rail tank cars. Some large chlorine production facilities have larger storage tanks (Chlorine Institute, 1999). If the pressure vessel containing liquid chlorine at ambient temperature is ruptured, a portion of the liquid chlorine will flash vaporize. Much of the remaining liquid chlorine is aerosolized—broken into droplets dispersed in the air—and then evaporates or "rains" back down to the ground (Barrett, 2009). If a cryogenic—low-temperature—chlorine tank is ruptured, the liquid chlorine does not vaporize, but instead leaks out as a pool of boiling chlorine liquid that generates a cloud of chlorine vapor (Barrett, 2009). Chlorine gas is heavier than air and will pool in low-lying areas.

[3] Because humans have rarely been exposed to lethal concentrations of chlorine gas, and when they are exposed the exact concentration is rarely recorded, there is a great deal of uncertainty about the threshold for lethal damages. Other sources cite lower harmful and lethal concentrations. Withers and Lees estimate that a 30-minute exposure to 210 ppm would be lethal to 50 percent of the population (Withers and Lees, 1985, cited by National Research Council Subcommittee on Acute Exposure Guideline Levels, 2004).

**Table 3.1**
**Health Effects of Chlorine Gas by Parts per Million (ppm)**

| Chlorine Concentration | Effect on Humans |
|---|---|
| 0.2–0.4 ppm | Odor threshold |
| 1–3 ppm | Mild, mucous membrane irritation |
| 10 ppm | Immediately dangerous to life and health (IDLH)[a] |
| 5–15 ppm | Moderate irritation of the respiratory tract |
| 30 ppm | Immediate chest pain, vomiting, dyspnea, and cough |
| 40–60 ppm | Toxic pneumonitis and pulmonary edema |
| 430 ppm | Lethal over 30 minutes |
| 1,000 ppm | Fatal within a few minutes |

SOURCE: Chlorine Institute, 1999.

[a] In conditions above the IDLH level, appropriate protective equipment is recommended before entering the area (Chlorine Institute, 1999).

perse (with the concentration dropping below the level at which it is a serious hazard), removing the threat to additional downwind sites.

Though an actual incident could have a linear progression like this, the specific conditions could produce very different geographical arrangements of threatened and eventually affected sites. Also, Figure 3.1 includes only those sites actually affected by the hazardous materials (hazmat) release. Other proximal sites could be affected in different ways (e.g., because individuals are evacuated from harms' way into a nearby site or adjacent community), which could result in different emergency response requirements and hazard conditions.

## Considering the Capabilities and Requirements for Responding to a Chlorine Release

Based on the incident schematic in Figure 3.1, there are three general classes of response options for a chlorine incident, summarized in Figure 3.2:

1. *Action at the Source Site.* If it is possible to contain release of the material at the source site, then the overall scale of the incident can be contained and (theoretically) the number of threatened sites with the potential to be affected reduced.[4,5]

---

[4] In most circumstances, the source site is also the first affected site, meaning that there will likely be victims there requiring assistance, as described below.

[5] At all sites, perimeter control as a method of preventing the exposure of additional unexposed individuals can contribute to limiting the numbers of casualties from a release.

**Figure 3.1**
**Schematic of the Time Evolution of a Chlorine Release**

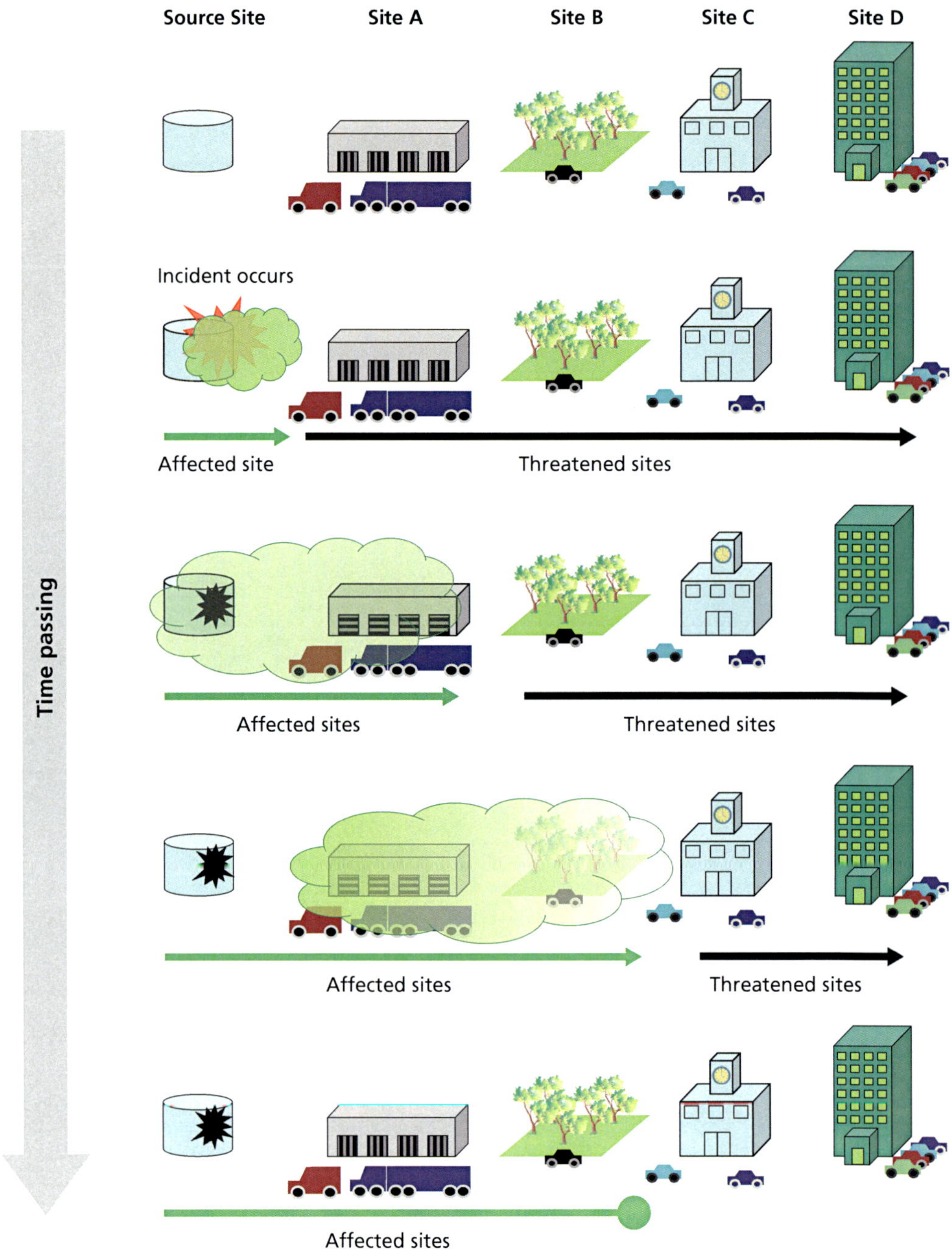


RAND *MG994-3.1*

**Figure 3.2**
**Response Options at Source, Affected and Threatened Sites**

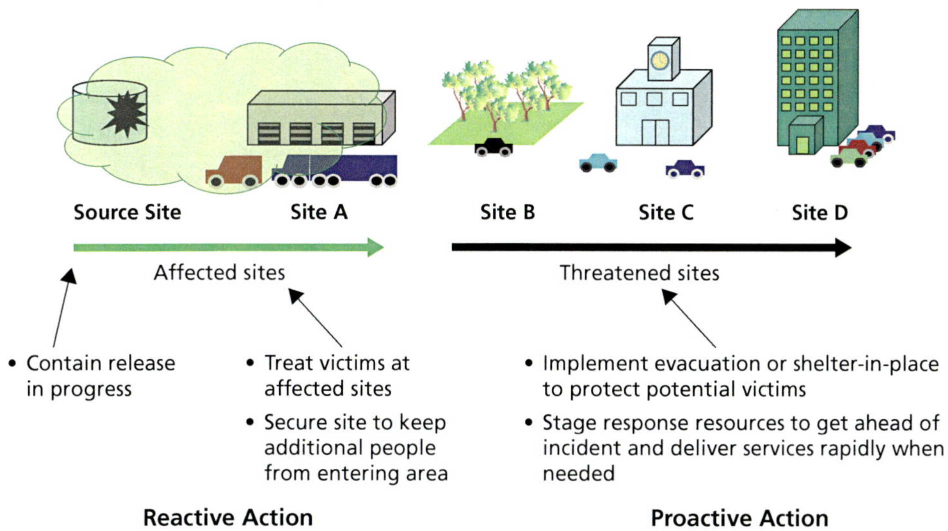


RAND *MG994-3.2*

2. *Action at Threatened Sites.* If the opportunity exists to get ahead of the incident and take proactive action at threatened sites, the potential consequences of the incident can be reduced.
   – Actions such as evacuation or shelter-in-place interventions at each site would shield or remove part of the potential victim population out of harm's way.
   – More-limited proactive actions, including predeployment of response resources in advance of needs, could make it possible to rapidly intervene when needed, since treatment after chlorine exposure is time-sensitive.
3. *Action at Affected Sites.* Once sites are affected, intervention is focused on keeping exposed individuals from becoming casualties. Actions include:
   – *Assisting exposed individuals to leave the cloud rapidly.* Fast rescue efforts attempt to keep their dose below the point at which they will be injured.
   – *Decontamination, if necessary.* Though exposure to chlorine gas does not usually require victims to be decontaminated, some individuals might require it. Structures or material exposed to sufficient chlorine to cause damage might require decontamination.[6]
   – *Treatment of "lightly injured" individuals at the scene.* Action to prevent individuals who were exposed but are readily treatable from progressing to more serious injury.

---

[6] These decontamination activities would likely be done during the less time-compressed recovery phases of an incident rather than the response phase.

- *Stabilization and transport of seriously injured.* More significant medical action (including movement to hospitals) of individuals at serious risk.

In framing the desired outcomes of response actions to chlorine spills, we have focused on the time-critical elements of response related to preventing or treating victim injuries. It is true that large chlorine releases can produce physical damage through the corrosive action of the chemical on plants and some types of physical infrastructure. However, with the exception of attempting to contain the release in progress, which would limit both casualties and damage,[7] many other actions taken to address these damages—notably decontamination and remediation efforts—are more likely to take place in the less time-constrained recovery phases of response after the life-safety components of response are complete. This is in contrast to some other types of incidents (e.g., wildland or other major fires), where the time scales of life-safety and property protection are more similar. As a result, in our analysis of response, we have framed the goal of action as attempting to cut the number of victims of the incident by (1) preventing exposure in the first place ("depleting the reservoir of potential victims") and (2) providing treatment to exposed individuals fast enough that their injuries do not become serious or fatal.[8]

For affected individuals, chlorine exposure is treated by removing the victim from the contaminated area and giving oxygen as soon as possible (Chlorine Institute, 1999). While exposure to gaseous chlorine does not require decontamination of victims, responders may decontaminate victims anyway if there is any uncertainty regarding what chemical victims were exposed to or if they are unsure whether the victim was exposed to liquid chlorine (Houghton, 2004).[9] Contact with liquid chlorine causes a freeze burn, with severity depending on the duration of exposure. Since liquid chlorine quickly vaporizes, exposure to liquid chlorine would only be a potential concern for individuals at the source of the leak (Chlorine Institute, 1999). However, chlorine gas can also combine with water vapor in the air and remain as a corrosive agent on clothing and equipment (Sanders, 2006). Local Fire Department, Surgeon General of the U.S. Army, and National Fire Protection Association guidelines all recommend that "When in doubt about contamination, decontaminate all involved personnel" before entering hospitals or receiving emergency care (Houghton, 2004).

---

[7] We did not address containment of the release, in an effort to simplify our analysis. This simplification is discussed in greater depth in Chapter Four, which provides more detail on the specifics of our analysis.

[8] This mission could be viewed in a simplified way as there being a theoretical number of possible victims from a release of a given size in a given area (based on population nearby, likely direction of transport, etc.), with response actions seeking to reduce the actual number of victims below that theoretical number through a variety of approaches.

[9] Decontamination of victims exposed to hazmat involves removal of contaminated clothing and washing of victims with water and potentially other materials (detergents, appropriate neutralizing agents) to remove or render residual hazmat harmless.

As Figure 3.1 suggests, there is a strong time dependence for the ability to take particular response actions. With the initiation of a release, an "incident clock" starts and advances as the release progresses and the cloud moves. Over time, the window for action at any given site will gradually close from the top of the list downward. Once the release has gone to completion, the opportunity to partially contain the material (and cut the scale of the incident) is gone. As threatened sites become affected sites with the movement of the cloud, the opportunity to take proactive rather than reactive action is lost. And, at sites exposed to the gas, the dose-response dynamic of hazmat exposure means that response action to meet individual victims' needs also has a strong time dependency. As exposure accumulates, exposed individuals will die, and the opportunity for response to reduce the impact of the event will be lost.

CHAPTER FOUR

# A Simplified Model of an Emergency Response to a Chlorine Release

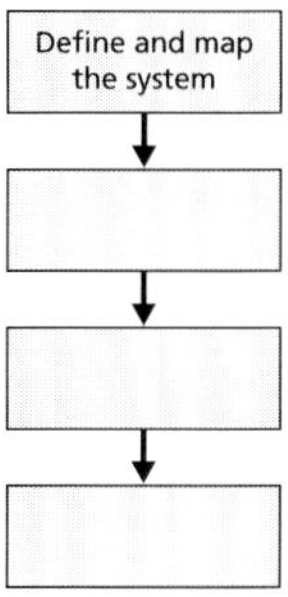


The first step of the FMECA analysis that we are adapting to examine preparedness for large-scale response operations is to define the individual parts of the system that are being assessed and identify their functions. Doing so requires building a block diagram of the system elements and their linkages to one another, and articulating what it means for the pieces of the system to work well. This chapter does both these things for the chlorine response scenario described in Chapter Three.

To address the requirements of an emergency situation, response organizations deliver capabilities that can accomplish the necessary tasks. To support national preparedness planning, the DHS defined a standardized set of response capabilities in the TCL (DHS, 2007b). Those capabilities range from incident-specific (e.g., Fire Incident Response Support) to more general (e.g., Emergency Operations Center Management). Such frameworks of capabilities provide a good starting point for identifying the key elements of the response system needed for a specific incident type.[1]

But to do a *reliability* analysis, more is needed than a recitation of the capabilities relevant to the particular scenario at hand. The architecture of how those capabilities fit together is also required. For example, a successful response must include both incident management and capabilities to treat victims. But the ability to do the latter depends on the performance of the former.

While an ideal response system would have enough equipment, material, facilities, and personnel available to perform all relevant tasks, in real life, resources are frequently constrained and choices have to me made. Making the right trade-offs is part of limiting the effect on performance of the system of breakdowns, such as resource scarcity. To make the right trade-offs, it is useful to have a picture of the response system constituted to manage and deliver capabilities to make it possible to catalog the

---

[1] Appendix B provides a crosswalk between the model of a chlorine response discussed here and both the categories included in the NIMS (DHS, 2008b) and the capabilities from the TCL (DHS, 2007b).

sorts of breakdowns and problems that can occur in either individual parts or the links between them that will hurt the ability to respond.

In doing so, the goal is to lay out the response system and its parts at the right level of detail. The response system must be described at a high enough resolution that actions and circumstances that might affect the performance of each step can reasonably be identified (i.e., if the model is not detailed enough, then it may not be clear how a specific failure mode—e.g., communications delays caused by increased cellular phone traffic after the release—will affect the response overall). However, the response system must not be described in too much detail either—the more steps that must be examined, the more labor- and data-intensive any such assessment becomes, potentially limiting the usefulness of the approach for preparedness assessment.[2]

## Top-Level Structure of Our Model of a Chlorine Response

The first step in building our model of a chlorine response is to define the overall elements involved in such an operation. In the language of the TCL (DHS, 2007b), these elements are the capabilities required at the event, though the way we assembled the model breaks them out somewhat differently from the TCL to help simplify the particular analysis we are doing here. Our model is based on a number of sources from the policy and practitioner literature. Beyond drawing on the TCL as previously discussed, our sources included (1) response doctrinal publications, such as the NIMS (DHS, 2008b), relevant preparedness plans, and planning documents;[3] (2) guidance, doctrine, and standard operating procedures from the response practitioner and specialized technical literature;[4] (3) descriptions of response operations in after-action reports;[5] and

---

[2] In modeling a particular type of response, it is reasonable to expect that different jurisdictions or analysts might build models with some differences between them. For example, though the major elements of the response would be the same (e.g., incident command, medical treatment), there might be some differences in how the response plans in a specific area linked those elements together, or some elements might not be relevant given the nature of the threats and hazards in an area. For example, though we have included evacuation as a component of our model, it is reasonable to assume that there are areas facing risks of chlorine release where evacuation is sufficiently impractical that it is not a major part of their plans.

[3] For example, DHS, 2007b; World Health Organization, 2009; NFPA, 1997; U.S. Army, 2003.

[4] DeAtley et al., 2003; Boisvert, 2007; Phoenix Fire Department, n.d.; Argonne National Laboratory, 2001; International Association of Fire Chiefs, n.d.

[5] In the course of the study, team members reviewed a wide variety of response after action reports, covering incident types that included, but were not limited to, chlorine and other hazmat release incidents. AARs reviewed included relevant materials from DHS's Lessons Learned Information System (LLIS), AARs made publicly available by response organizations on the Internet, and AARs from previous research efforts undertaken at RAND.

(4) academic studies identifying particular response functions, architectures, or simulations of response operations.[6]

Given the need for reliability analysis to address incidents of different sizes, we framed our response model in a way that makes it applicable to incidents of various scales. For example, while evacuating areas would be very different prospects for small versus large chlorine releases, we have designed our model to accommodate differing scales of that function. A simplified schematic of our response model is included in Figure 4.1, which lays out the main elements of the model focused on addressing the requirements of an incident. This schematic includes only the overarching categories of model components, and not the detailed steps within each of the functions, which we discuss next.

**Figure 4.1**
**General Architecture of Our Model of a Chlorine Response Operation**

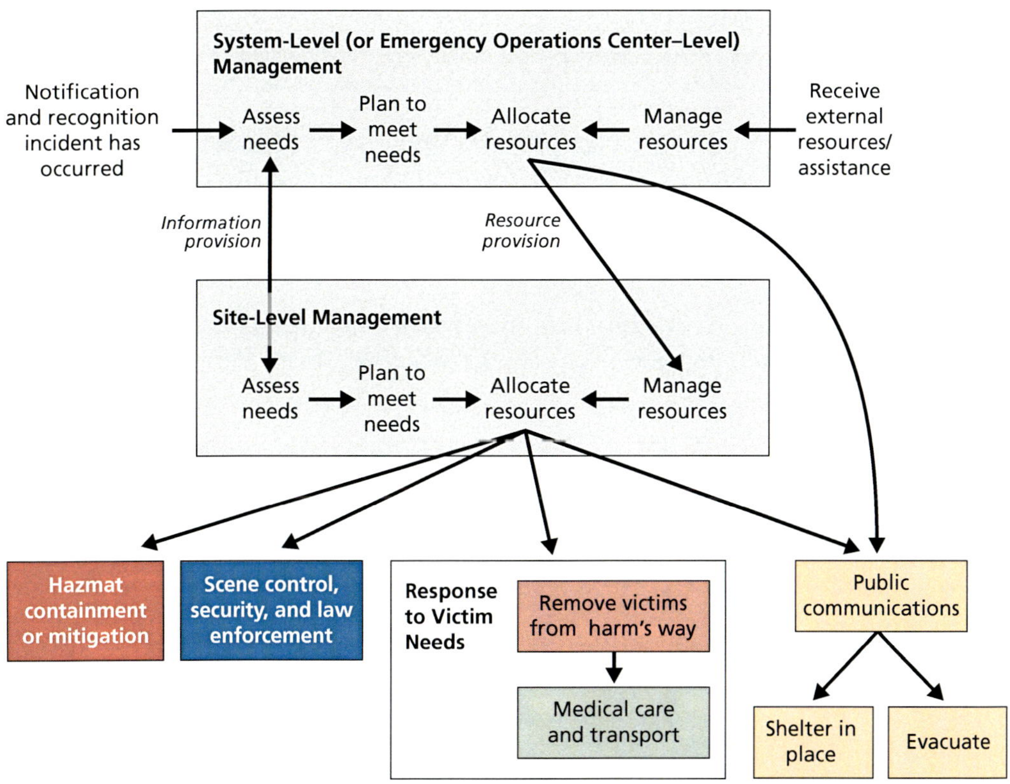


NOTES: Our model also includes a responder safety and health function, described later in this chapter. We do not show this function here for the sake of clarity, since it links to all the other functions.
RAND *MG994-4.1*

---

[6] Jackson et al., 2004; Jackson et al., 2002; Houghton, 2004; Barrett, 2009; Institute of Medicine, 1999; Raber et al., 2002; Byers et al., 2008.

To cover both large and small incidents, the model includes both system-level incident management[7] and on-scene, site-level management operations. Particularly small incidents might not have system-level management and instead be managed only at the incident scene, at which point the additional layer might not be relevant. Particularly large incidents might have operations at multiple incident scenes, keeping the upper layer and replicating the site-level incident command structure as area commands in different affected locations. The functions of command at whatever level are to process information and build situational awareness to identify response requirements, plan to meet them, coordinate the resources to do so, and allocate those resources to tasks. What resource coordination involves would similarly vary by incident scale, with large incidents involving resources coming to an incident through mutual aid from many more response organizations (creating a potentially more complex coordination effort) than for smaller events.

The response tasks we included in our model fall into the following general categories:

- **Hazardous Materials Containment or Mitigation at Source.** This category covers efforts to reduce the size of the release itself by stopping the release in progress, containing, and disposing of the hazmat released.
- **Scene Control, Security, and Law Enforcement Function.** For intentional releases, law enforcement action will involve investigation and crime scene operations. All events would likely involve perimeter control (which might be performed by police or other responders, such as fire service or hazmat) and potentially other scene security functions.
- **Response to Victim Needs.** This involves helping people out of the chlorine cloud if applicable (requiring appropriately equipped responders to operate in a hazardous environment), treatment of victims at the scene, and transport of the seriously injured to medical facilities.
- **Public Communications Functions.** Beyond simply ensuring that accurate information about an incident is available to the media and the public, in incidents such as chlorine release events, public communications can be part of the response strategy. For interventions such as evacuation or sheltering in place, the public must be told what to do in order to protect themselves. In our model, that direction can come either from the EOC level of the response (e.g., for wide evacuations in large incidents) or at the scene (e.g., for smaller incidents for which evacuation or shelter-in-place is more tactical or local).

---

7 In accordance with the terminology used in the TCL (DHS, 2007b), in Figure 4.1 we have labeled our "system-level management" as emergency operations center (EOC) management, though we note that level of management might not necessarily be in an EOC as such, and could take place in one of the purpose-built organizations/structures specified in the NIMS (DHS, 2008b) for large-scale incidents. In the text, we use the terms *EOC management* and *system-level management* interchangeably.

For our analysis, we made a set of assumptions and simplifications to limit scope and complexity. First, we focused on actions taken after release of the chlorine, irrespective of the cause of the release. As a result, we neglected specifics of law enforcement actions and explosive ordnance disposal—both of which would be key in a terrorist-initiated release—other than how carrying out those activities might mean that law enforcement officers were unavailable for other response tasks. Second, we assumed that the release of chlorine in our scenario was such that efforts to halt the release at the source were impossible or of only modest importance for the outcome of the response action. As a result, the mechanics of hazmat mitigation at the source site are not addressed. Finally, we stopped our analysis at the point at which victims were transported from the scene to medical facilities; beyond the "hand-off" of patients from EMS to hospital care, we did not address their subsequent treatment. For the same reason, our analysis did not address fatality management.

## Detailed Discussion of Two Exemplary Model Components

To carry out a reliability analysis, the individual steps required to actually deliver the capabilities identified in Figure 4.1 have to be laid out. Furthermore, the interconnections between different parts—how they depend on one another for response success—must be identified. In contrast to the simple, linear example in Chapter Two, a real response operation's elements are interconnected and interdependent. Delivering medical care to a victim of chlorine exposure requires not just having the resources needed for medical care and transport, but also an incident management structure that can allocate those resources effectively, information and situational awareness to know where victims are and what they need, perimeter control to keep concerned or curious members of the public from interfering with treatment activities or becoming additional victims themselves, and so on. Within each of the categories discussed above, our model therefore drills down to a higher level of detail, breaking down each activity into individual components of greater specificity.

That detailed model is shown as a thumbnail image in Figure 4.2; a larger version is included as a fold-out insert in printed copies of this document and is available for download as a PDF on the RAND website.[8] The remainder of this section walks through two of the model's six branches—system-level incident management and response to victim needs—in greater detail. (We walk through the other four branches in Appendix C.)

Identifying what might go wrong in a response (our failure mode analysis)—and, more importantly, the potential effect of individual things going wrong on performance—requires that the model have a clear articulation of what it means for

8 http://www.rand.org/pubs/monographs/MG994/

**Figure 4.2**
**Thumbnail Image of the Chlorine Response Operation Model**

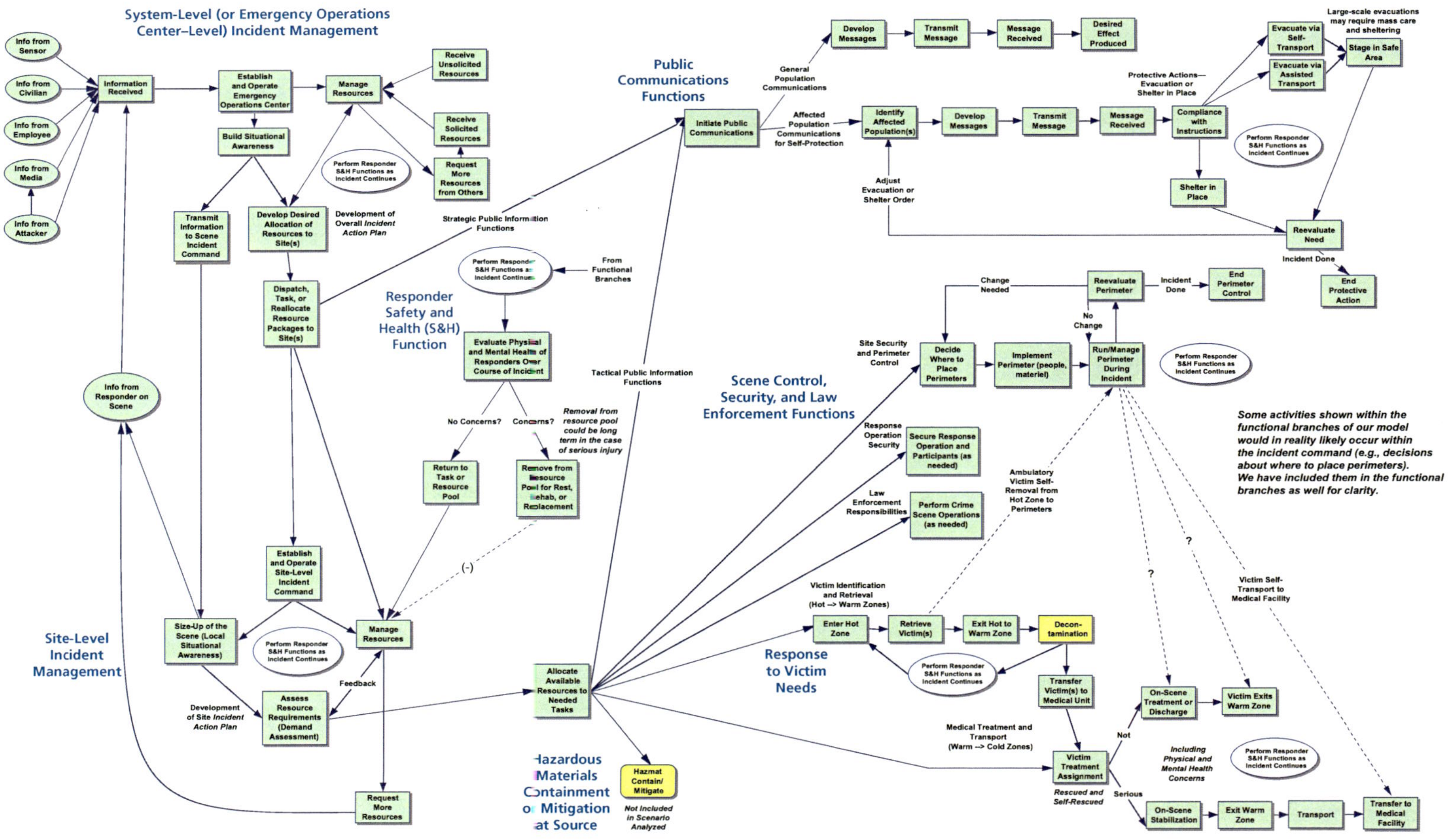


NOTES: A larger version of this figure is available as a fold-out insert included with printed copies of this document. The larger version is also available for download as a separate PDF at http://www.rand.org/pubs/monographs/MG994/.

RAND MG994-4.2

each step of the operation to work well. We therefore also describe in a general way what it means for each of the steps to function effectively (what we have called "measures of merit" for the steps). These measures provide the basis for identifying individual failure modes that threaten performance.[9]

In considering measures of merit, we have attempted to be minimalist. For example, in considering the effect of information quality on response management, not every step of our model that involves receiving information has a measure related to the accuracy of that information. Instead, later steps relating to the overall picture of the incident built by the incident managers have quality and accuracy measures associated with them, since it is at that point that information quality problems translate directly into deleterious effects on the response. Put another way, it doesn't matter if a mix of accurate and inaccurate information flows into the incident command if the command is effective in identifying and discarding the inaccurate data (because of the expertise of the individuals involved, the processes applied, redundant information sources for cross-checking, etc.). Part of the goal in keeping the set of measures of merit to a minimum is to simplify later phases of analysis as much as possible.

We have attempted to structure the model so it can be considered both statically—i.e., as a model of the initial decisions that are made and actions taken in an incident—and dynamically, so that we can consider both initial response actions and how those actions might be adjusted over time. As a result, in some cases our measures of merit (e.g., the time elapsed in a step) can be viewed as applying to initial actions (e.g., setting up incident command) or as applying to operations that are adjusted to respond to changing circumstances. The consequences of a failure mode would differ depending on whether it occurred initially (an initiation response termination or capability reduction through delay) or later (a random termination or capability-reduction failure).

## System-Level Incident Management

The system or EOC level of the incident command represents the management actions taken by response organizations away from the scene of the release. Our model of the system-level management is shown in Figure 4.3. For smaller incidents, such actions could be very modest (or even nonexistent), but for large-scale releases there could be significant coordination required at this level for allocation of response resources among tasks and sites. The steps in the model, moving generally from left to right are as follows:

---

9 Our development of these measures of merit was informed by a review of the TCL (DHS, 2007b) and the metrics and measures for different response functions included throughout that document. To draw on the TCL as a source for thinking through measures of merit for different response capabilities or functions, we reviewed *all* of the metrics included in the document and sorted them based on their relevance to the different parts of our model. As a result, in considering our measures for incident management, we drew on not only the metrics in the TCL for that function in particular, but also those discussed in other capabilities that related to such tasks as the gathering and analysis of information, resource allocation, and resource management.

**Figure 4.3**
**System-Level Incident Management Components of the Chlorine Response Operation Model**

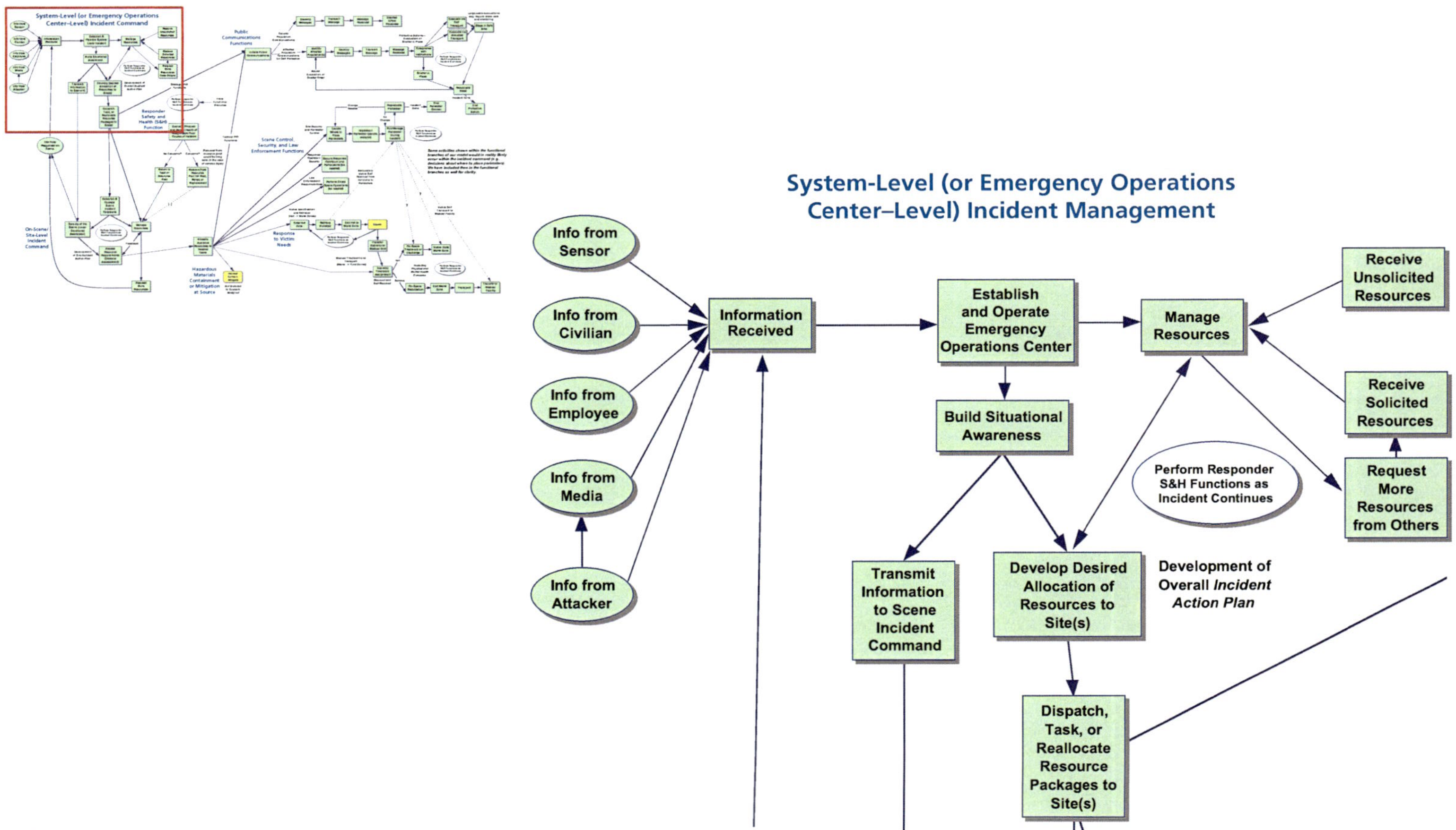


RAND *MG994-4.3*

- **Receipt of Information About the Incident and Its Characteristics.** To manage an incident, the incident management system (IMS) needs information about it. Considering measures of merit, for the initial notification that a chlorine release has occurred, it is reasonable to assume that response systems will be notified by some means, but the central question is how long it will take for that notification to be received. For later phases of an incident, when more detailed information is coming into the system-level incident command, or EOC, from the scene(s) of response activity, both (1) the time that that it takes to communicate relevant changes in the situation that require action at the system level and (2) the accuracy/completeness of information transmitted become more central as parallel measures of merit.
- **Establishing and Operating System-Level Incident Command (the EOC).** The IMS itself is the physical and human nerve center assessing information, planning, and executing management tasks. For doing so, the main measure of merit is whether the incident command/EOC is functioning such that it can execute its planning, assessment, prioritization, and allocation tasks effectively.[10] The centrality of the incident command/EOC in guiding operations means that sufficient problems in its functioning could affect many other parts of the response.
- **Building Situational Awareness.** To guide its top-level coordination of response activities, the IMS requires situational awareness of the incident—the sites that are affected, the types of requirements at those sites (number of victims, severity of exposure, etc.). We include in our notion of situational awareness both a picture of the current incident status (essentially, current knowledge of the affected sites) and a projection of the future incident status, including a reasonable prediction of which sites are threatened by the release. This step has a time-elapsed measure of merit (since time spent in incident command initially delays the response, and, later on, delays in updating situational awareness could hurt effectiveness.)[11] However, it has an accuracy component as well. For the picture of the current incident status, the two parameters are that (1) all sites that have been affected are correctly identified as affected (and reasonable estimates are made for the needs at those sites) and (2) no sites that have *not* been affected are mistakenly identified as affected. For the projection of the future course of the incident, the measures are essentially the same, but the sites at issue are those that will be affected as the incident progresses. If sites are left out, needs will go unmet at those sites. If sites

[10] A delay in setting up incident command at all (e.g., effective management is not established for tens of minutes or hours into response activities) would be viewed as a complete quality breakdown for the period involved.

[11] The time required to build situational awareness will be affected by earlier steps in the model. For example, if the initial reports of the incident came from an employee at the site of the chlorine release and could provide information on the total volume, speed of release, etc., in addition to the fact that the incident had occurred, the picture of the incident and the projection of its likely course could be built more quickly.

are misidentified as affected or threatened, resources will be wasted by allocating them to sites where needs do not exist or will not arise in the future.[12]

- **Developing Desired Resource Allocation for Sites.** Given the picture of needs and projected progression of the incident, resources requirements are developed. This process is essentially building the incident action plan (IAP) for the response operation at the system level, where the understanding of incident needs and knowledge of response resources available (see below) are combined to arrive at a desired allocation of resources to sites and tasks. This step also has a time-elapsed measure of merit (since time spent developing the IAP might slow response or, later in an operation, slow adjustment of the response operation to new information). The other measure of merit is the appropriateness of the match made between available resources and assumed requirements (e.g., for the number of victims thought to be at a particular site, an appropriate number of medical and other responders is allocated to treat them).

  If available resources exceed the perceived needs of the incident, this step may simply involve matching a subset of actual response units to those needs. In a situation where the requirements are closer to the performance limits of available resources, this step requires deliberate allocation of those resources among sites and tasks.[13] For situations of resource scarcity, the ideal would be to optimize the allocation for the greatest reduction in harm that can be "bought" based on the resource package at hand.
- **Managing Resources.** The other core activity of the IMS at the system level is the coordination of resources. One of the core functions of the NIMS (DHS, 2008b) in particular and incident command/management systems in general is that resources from multiple responding organizations can be integrated effectively to perform together in a unified and coordinated way. That coordination and management includes maintaining an inventory of resources (i.e., management of the *knowledge* of what resources are available) and *physical* resource management (i.e., staging for response units or warehousing for other necessary supplies). The central measure of merit for this function is the "effective size" of the coordinated resource pool that is available at the response and how the size of that pool evolves over the timeline of the response. That is, if resource coordination and management are very effective, the response will be able to efficiently utilize all the resources that participating organizations bring to the operation because managers know the units are there and what they can do, and can call

12 As we will discuss later, the seriousness of this type of inaccuracy is linked to how the scale of the incident corresponds to the maximum capacity of the response organization(s) involved (see Figure 2.3). If the organization is operating well below its maximum threshold of capability, there may be enough slack resources to serve all sites—both those correctly identified and misidentified.

13 For example, at the limit of the maximum theoretical performance of a response system, this resource allocation process would have to use each resource to its maximum efficiency to deliver to its full potential.

on them when needed. If resource management is less effective, either resources may become "practically unavailable" (e.g., even though they are there, the IMS doesn't know it because of a breakdown in personnel tracking or linkage of units into the IMS) or their availability may be delayed.

- **Requesting Additional Resources.** When a response operation is short of resources, then another core function of the system-level incident command/EOC is to request and integrate resources from outside sources. Resources that might be called on include mutual aid from other response organizations and help from volunteer organizations or even individuals. In our model, resources from both these sources would fall into the "receive solicited resources" box. Depending on the nature of an incident, unsolicited assistance might also be offered.[14] For all sources of outside assistance, the measure of merit is whether aid can be received and coordinated into the pool of resources at the incident such that it represents a net increase in response capabilities.[15] The linkage of additional resources to the IMS so that they are integrated into inventories and response planning is a key component of this process. This step also has a time dimension, since outside assistance will only be of value if it can arrive quickly enough to take productive action. Given the relatively rapid timeline of chlorine release incidents (which evolve over hours in most cases, rather than days or weeks like hurricanes, earthquakes, or major fires), assistance from organizations at significant distance from the incident site would likely be less relevant than that from nearby sources.
- **Transmitting Information to the Site-Level Incident Command.** In our model, the system-level IMS has two major connections with the incident scene. The first is its role in passing situational awareness information to the scene to inform decisionmaking and management at the site level. In our model, we have this communications function linking situational awareness at the system level down to the incident scene. The central measures of merit for this function would be both the accuracy and relevance of the information provided by the EOC-level IMS to the scene and its timeliness (e.g., how quickly data on changes in the incident are passed to those who can act on them).[16]

---

[14] There is an extensive literature on the effects of unsolicited assistance that is not coordinated with response operations. Such assistance from untrained volunteers and even from trained individuals (e.g., so-called "free-lancing" responders) can be significantly disruptive.

[15] To illustrate with an extreme example, if untrained volunteers responded to an incident such that one local responder had to accompany each volunteer to make it possible for them to work, there would be no net increase in resources. One of the fundamental principles of the incident command system and NIMS (DHS, 2008b) is that the common operating standards and practices will make it easier for outside resources to seamlessly integrate into a response operation to augment capability.

[16] In this case, timeliness could be viewed as integral to information quality, since information would become less relevant as the incident continued to evolve.

- **Dispatch, Task, or Reallocate Resources to Sites.** Our second core connection between the system and site levels is the provision of resources—the centralized dispatch of response forces from the EOC level to the site or sites affected by the incident based on the resource allocation previously described. As with the previous steps, this step has a time component to its effectiveness, since delays would cut into the ability of those resources to act. Its central measure of merit would be accurate and successful dispatch of resources to the scene(s) where they are needed.
- **Perform Responder Safety and Health Management Functions.** Throughout our model, performance of responder safety and health management functions is shown as a unique activity that may or may not be not directly connected to other pieces of the system model. This is intended to designate that this command staff function (e.g., as it is positioned in NIMS) is performed for all of the responders involved in incident command, not just those involved in "functional" response activities, such as victim retrieval and treatment. The oval in the model, shown in white, designates a general link to the responder safety and health function (described in Appendix C).

### Response to Victims' Needs

In the functional branch where our model covers actions directly taken to meet victims' needs, we have two categories of activities: (1) victim location and retrieval—essentially the entry by appropriately protected responders into potentially still-hazardous environments to find victims, remove them from the hazardous environment, and transfer them to medical care—and (2) medical treatment and transport—involving both the on-scene treatment and release of minimally injured individuals and the more extensive stabilization, transport, and transfer to hospitals of more seriously affected individuals. Based on general hazmat operation guidelines, the first category roughly corresponds to operations in the hot (or formerly hot) zone and transition to the warm zone. The second covers actions in the warm and transition to the cold zones of operation. These are diagrammed in Figure 4.4.

#### *Victim Identification and Retrieval*

- **Enter Hot Zone.** Finding victims of a chlorine release, particularly for a release that is still in progress or early enough in the incident that the environment is still hazardous, requires entering that environment. This requires appropriately outfitted and trained responders who can operate and provide assistance under those circumstances.
- **Retrieve Victim(s).** Once found, victims must be removed from the hazardous environment to halt their exposure and get them to an environment where they

**Figure 4.4**
**Response to Victim Needs Components of the Chlorine Response Operation Model**

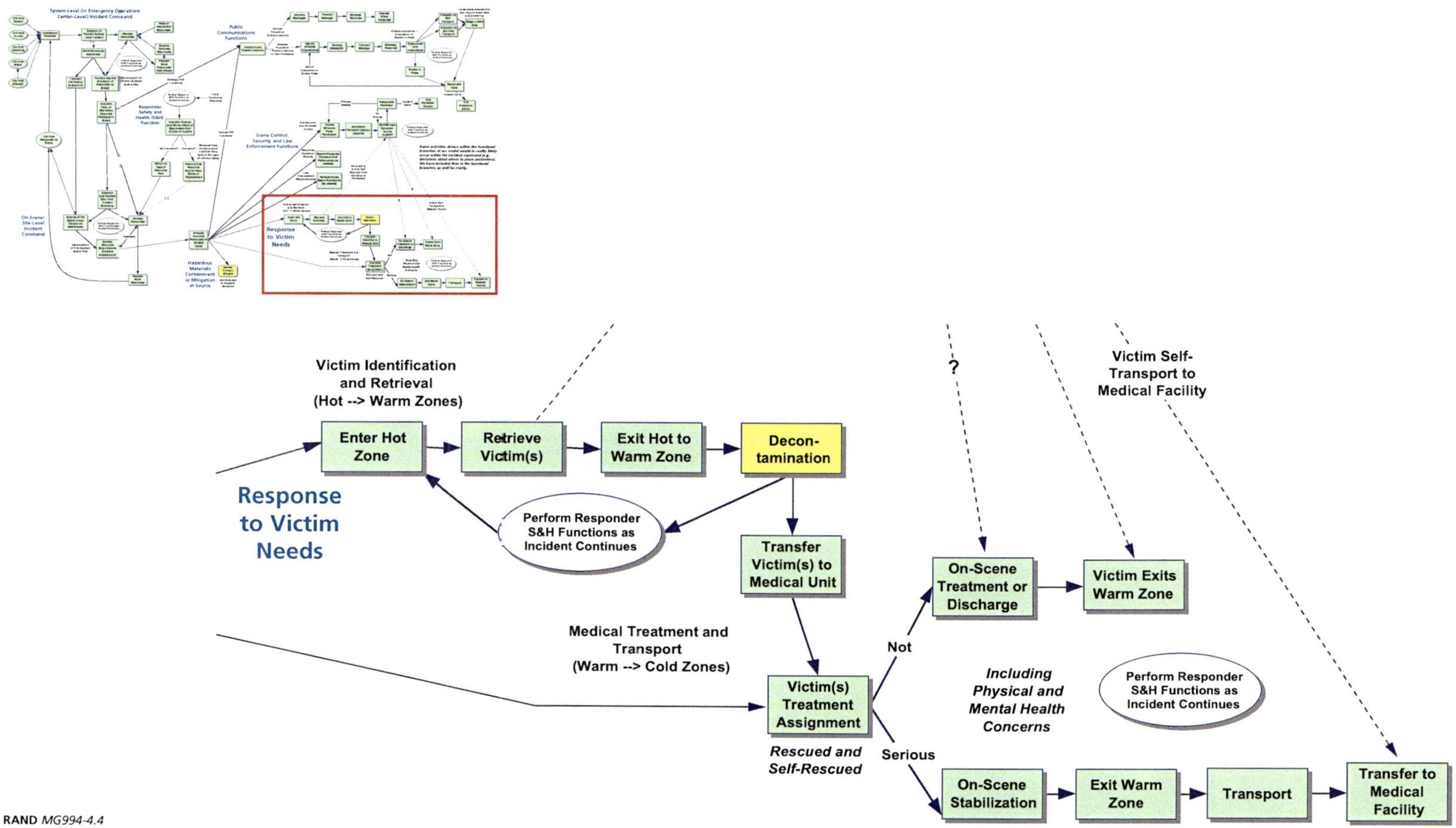


RAND *MG994-4.4*

can be safely aided. Individual responders within the hazard zone will be able to retrieve victims one or more at a time, depending on whether those victims are ambulatory or must be carried to safety.

- **Exit Hot to Warm Zone.** Once retrieved, victims are transferred from the hot zone to the warm zone.
- **Decontamination.** Though decontamination is generally not necessary for individuals exposed to chlorine (unless topically exposed to liquefied chlorine), it is included both for completeness and because it is a function that might be implemented as a general precaution early in the response, before the nature of the incident is clear, or at the release site, where exposure to liquid chlorine would be more likely.

With respect to measures of merit, in this case, we believe it makes the most sense to have an overall measure for these four steps rather than breaking them down individually—though individual failure modes could obviously affect one of these steps but not others. Our overall measure is the fraction of affected victims removed from the contaminated zone fast enough to prevent permanent morbidity or mortality. As before, this has an embedded time dimension, since faster removal will reduce victims' exposure.

- **Transfer Victim(s) to Medical Unit.** In our model, we include a step for the transfer of victims from responders retrieving them to medical units trained to evaluate, treat, and potentially transport them. This might be a transfer from hazmat or fire service personnel (since those specialties would presumably have the equipment required to operate in a chlorine-contaminated environment) to EMS personnel. In some cases, with responders cross-trained or involved in a very integrated effort, this handoff might not be required. The fundamental measure of merit for this step is the average time elapsed for patients between their retrieval and initial treatment assignment, since that time could result in more serious injury or fatality after chlorine exposure.
- **Perform Responder Safety and Health Management Functions.** As above, responder safety and health is included as a "linking function" to the specific branch of the model that involves assessing responder safety issues.

#### *Medical Treatment and Transport*

In the model, the medical treatment and transport step includes two branches: one for serious injuries and another for non-serious injuries (or uninjured individuals who require only discharge). The common features for those two paths are

- **Victim Treatment Assignment.** The first step in the process is to assign victims to treatment, to determine whether a victim requires more major medical intervention (including transport to a medical facility) or can be treated on-scene and released. The main measure of merit for this decision step is the average time that it takes to assess a patient and initiate appropriate care. The accuracy of assignment is also clearly important—i.e., correctly identifying those who can be treated on-scene versus those who need more extensive on-scene stabilization and transport—though, in practice, miscategorization could reduce to an additional delay if the mistake was recognized in the course of either on-scene treatment or stabilization for transport.

- **Perform Responder Safety and Health Management Functions.** As above, responder safety and health is included as a "linking function" to the specific branch of the model that involves assessing responder safety issues.

Below, we describe the steps for the two branches.

*Minor Injury Branch*

- **On-Scene Treatment and Discharge from the Warm Zone.** Our model includes two elements on the path for treatment of minor injuries: on-scene treatment and then discharge of the patient out of the warm zone (an "end state" for action of the response system with respect to that individual). The measure of merit for this step is the fraction of relevant victims (i.e., those with injuries that are not immediately serious) who are treated fast enough to obviate the need for additional medical care. For example, if delays in treatment meant that some less serious injuries developed into medical emergencies requiring more extensive care, those cases would be viewed as a reduction in response effectiveness on this branch.

*Serious Injury Branch*

- **On-Scene Stabilization.** For serious casualties, our model assumes that some medical intervention would be necessary on-scene to stabilize the patients for transport. For chlorine injuries, this could include respiratory support, among other emergency medical treatments.
- **Exit Warm Zone.** Exit from the warm zone for seriously injured patients would presumably require assistance from the EMS responders involved in transporting them to a medical facility.
- **Transport.** Our transport step includes all the actions required to for patients to be moved from the scene, in a vehicle appropriate for their medical requirements, to a hospital or other medical facility prepared to receive them and provide further treatment.
- **Transfer to Medical Facility.** As discussed above, a transfer step to a receiving medical facility is included as the end point of the branch, signifying the transfer of the patient from responders involved in the incident response to the health care system.

As was the case with the initial steps of victim location and retrieval, we believe that measures of merit for the serious injury branch of medical injury and support are best framed as a single measure for the process overall, though different types of failure modes might affect only one of the four steps. For this set of steps, the overall measure of merit is the fraction of seriously injured victims who are stabilized and transported to receiving facilities fast enough to prevent permanent morbidity and/or mortality.[17]

## Discussion

Assessing the reliability of a response system requires articulating not just the pieces of the response—the capabilities that are needed and the amounts of them that are required—but also how these pieces fit together. Some of those capabilities—i.e., the ability to carry out incident management—are inherent parts of building that system. Other capabilities—such as emergency medical treatment—are the outcomes that system delivers to people in need. The linkages among the parts of the system determine how problems in one will—or will not—affect its performance in other ways.

In putting together our model of an emergency response to a chlorine release, we have attempted to keep the individual elements of our model as simple as possible, since the subsequent steps of our examination—the analysis of the ways the various parts of that system can break down—will add another layer of complexity. As a result, some pieces of our model combine the efforts of multiple responders, or in some cases multiple response organizations (e.g., our situational awareness block could involve experts from multiple organizations involved in site and risk assessment).[18] However, at the level of aggregation of the model, the individual components provide a useful organizing structure for working through the various threats to system reliability and performance that we address in the next chapter.

---

[17] As discussed in the introduction, we end with transfer to a medical facility and do not address the additional issues of capacity and reliability, which apply to the functioning of the medical facility itself and its ability to provide surge treatment capacity for mass casualty incidents of various sizes, nor stocks of specialized medical supplies to provide treatment to the casualties of a chlorine incident (beyond the requirement that the transport elements of the response system deliver casualties to a medical facility that says it can receive them).

[18] It is also the case that the way we have structured our response model is different in some respects from other ways response operations and capabilities have been structured (e.g., the capabilities in TCL [DHS, 2007b]). This issue is discussed in Appendix B of this report.

CHAPTER FIVE

# Exploring What Can Go Wrong During a Chlorine Response Operation: Identifying Relevant Failure Modes

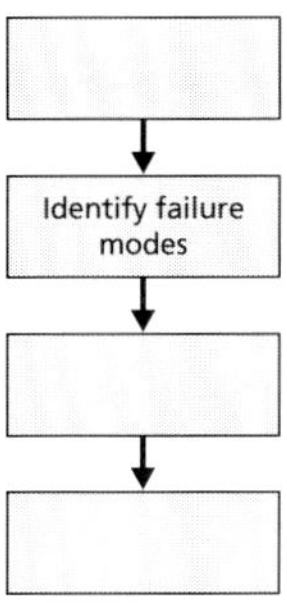


Having mapped out the response system and identified the linkages among different activities and components, we come to the second step in our adapted FMECA, which is to identify failure modes. Failure modes are "the observable manners in which a component fails" (Ebeling, 1997, p. 168). Identifying failure modes for an entire system requires systematically walking through "what could go wrong" at each point in the system that would observably affect response performance. But failure mode analysis is more than developing a list of potential problems that could get in the way of response.

In contrast to the example in Chapter Two, where a list of failure modes was presented as a complete package and the only task was to assign them to parts of the response model, breaking down a more complex system and determining *de novo* how the system could fail is a more complex and iterative process. The goal in doing so is to examine each of the parts of the system and develop a taxonomy of what events might affects its performance. Given a response operation, the process needs to aspire to be *comprehensive*—that is, to capture all significant failure modes—since ways that the system could breakdown that are not identified represent vulnerabilities to both the accuracy and validity of the analysis.

Although the results need to be comprehensive, the same concerns about complexity that affected the construction of the response model apply here. Including every possible thing that might happen, no matter how minor the probability, would quickly result in lists of failure modes so long that their completeness would crush their utility. As a result, a balance must again be struck, such that failure modes that are *known* to be significant (e.g., breakdowns in response vehicles limiting mobility) are included explicitly, while other events that *might* affect response are addressed by including an "other" category; if circumstances change and the chance of a previously negligible-probability event increases, that failure mode can be promoted out of "other" to be treated in its own right.

It is also the case that the ways that real systems break down are often more complicated than single, self-contained failure events that can be linked to the individual pieces of the response operation. In reality, sometimes the occurrence of a failure requires more than one thing to happen at once. For example, if a response has both a primary and backup communications system, both must fail at the same time to significantly affect the response. Some failure modes require the combination of a problem inherent in the system with an outside event to produce a breakdown. And, almost certainly, there is more than one thing that could go wrong that would produce a failure in a specific part of a response system: Incident command might be disrupted by communications breakdowns, but it could also have problems as a result of key staff being missing, all necessary organizations not being effectively linked into decisionmaking, and so on. Consequently, a failure mode analysis for any but the simplest of systems will result not in a list of stand-alone failures, but rather in a failure *tree*, laying out the relationships among different failure modes and how they—either singly or in combination—result in effects on system performance.

Figure 5.1 is such a tree, built from the examples discussed in the previous paragraph. Though there are multiple possible failures that could cause a breakdown (the three "tree branches" linked by an OR gate), for two of them there are multiple things that must occur (linked by AND gates) before a breakdown will happen.

**Figure 5.1**
**An Example Failure Tree**

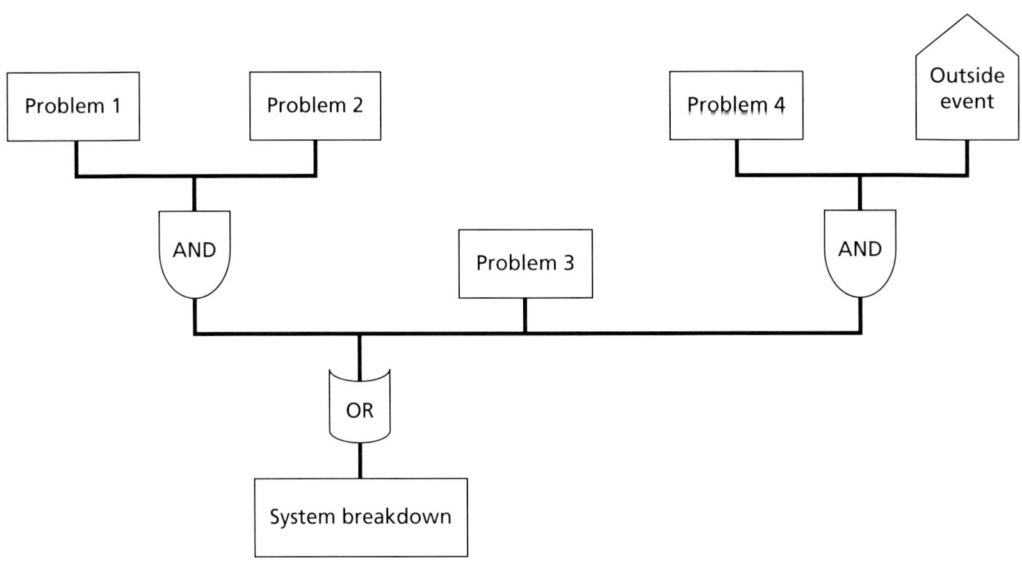


RAND *MG994-5.1*

## Building a Failure Tree for a Response Operation

Because a response operation—in our case, to a chlorine release—is a single complex system, a single failure tree could be constructed that included failures at each point in the system and described—through a complex web of AND, OR, and other linkages, how each individual failure event affected the performance of the system as a whole. Given the level of complexity involved in just building the model of the response system itself, it is clear that the result of such an effort would be highly complex.

One of the ways that FMECA analysis simplifies analysis of complicated systems is to allow them to be broken into pieces and the failure modes of each piece examined individually—while relating the consequences of failures of each piece to performance of the system overall. As a result, for our examination of response systems, we focus on building up failure trees not for the entire system, but for the individual functional parts of the system—i.e., subsets of the response model made up of one or more of the green boxes in Figure 4.2.[1] This both simplifies the analysis and eases presentation of the results, since separate component or element failure trees can be presented separately.

So how is this done? There are three basic steps to identifying and assigning failure modes to each part of our response model:

1. **Define the "system failures" for that portion of the response model.** This step essentially just frames the desired outcome of the element (the measures of merit discussed in the previous chapter) in the negative. For example, in the case of the "Information Received" block in the top left of the system diagram (Figure 4.2), the desired endpoint is information about the incident flowing into the system-level command. Failures would be bad information flowing in, good information not flowing in, or the delay of good information to the point where it was no longer valuable. For the purposes of this analysis, we generally consider failures in a given block—in this case bad information, missing information, or information delay—to have the same potential root causes, in an effort to simplify analysis as much as possible.[2]
2. **Define logical classes of failure modes that could produce system failure.** If a failure mode analysis of a system as complex as a response operation is going to be comprehensive enough to be useful, a defined process must be used to make sure the process systematically looks for and assesses all relevant failure modes. We did this by logically laying out, for each of the system-level failure modes, classes of failures that could produce the same end result. These classes

---

[1] We will discuss relating the effects of failures to system performance in the next chapter.

[2] We will discuss some of the negative consequences of this simplification in the next chapter, when we present our analysis of response AARs as a data source for response reliability analysis.

were defined by different types of intermediate failures that could produce the end system failure. In many cases, these classes were based on the main components that made up the response operation—plans and processes; technology and equipment; human resources—as well as external events (e.g., the response operation itself being disrupted by the chlorine cloud). In other cases, the logical breakdown was by steps in a process (e.g., request of mutual aid, receipt of that request, sending of resources, and arrival at the scene), where the classes were defined by what might go wrong with each step. In building classes of failure modes, we included a first layer of thinking about which classes could produce system failure on their own (and would therefore be linked by ORs in the final model) and which could only do so in combination with others (and would therefore be linked by ANDs).

3. **Identify the root failures that could result in system failures.** Given classes of failure modes, the last step is to identify possible root causes within each class that could produce system failure. Those events, which hereafter will be called basic or root failures, are systematically identified and added to the failure tree by AND or OR links to appropriately show how their occurrence will result in failure of the system component. Similarly to our development of the system model itself, our specific failure mode development for our chlorine response model drew on a variety of sources in the practitioner and academic literature. Basic failures are suggested by the variety of measures and metrics included in the TCL (DHS, 2007b), examination of AARs for past incidents that provide direct evidence of failure modes,[3] and the logic of doctrinal publications such as the NIMS (DHS, 2008b). As was the case in developing classes of failure modes, our goal was to be comprehensive but not fully exhaustive, to keep the failure trees to a workable level of complexity. Finally, every branch in the failure diagram is given an "other" failure mode to capture root causes that are not explicitly identified in the diagram.

When we actually carried out each of these steps for our chlorine response model, we identified two practical changes that were required from this idealized process for developing individual failure trees for each piece of the response model in isolation. Both can be viewed as adaptations needed when applying the FMECA technique to a complex human system, rather than a technical one.

---

[3] Our specific analysis of AARs is discussed in the next chapter. Our work with AARs meant that the analyses described in this chapter (identifying failure modes) and the next (collecting data from real-world sources on failures during responses) were done in an iterative manner. During the first half of the AAR review, we adjusted the failure modes based on the data being collected on actual failures from the AARs. By approximately midway through AAR review, we felt that the failure modes were sufficiently detailed for the purposes of this study, and so we froze them. We then recoded the AARs examined to that point based on the final failure mode taxonomy.

First, though the goal in FMECA is to treat each element of the system in isolation, the interdependencies that exist within a response operation do not make that entirely possible. In building out the root causes of failures in different parts of the system, in a number of cases performance depended closely enough on performance in other parts of the system that each part could not be viewed in an isolated way. For example, in considering whether mutual aid resources will be available to help assist in response operations, one of the classes of failure mode consists of reasons why a needed mutual aid request might not have been made in the first place. One such reason must be that the system-level incident command/EOC is not functioning properly. Since the functioning of incident command is addressed in another part of our response model, this essentially forges a link between the two parts of the failure tree, such that a failure in incident command is also included as a failure mode in the portion relating to request of mutual aid resources. As a result, although in our final analysis we sought to treat each part of the model as independently as possible, the end result was a set of separate failure trees wired together through their interconnections into something essentially representing a hybrid between a single overall failure tree and a family of separate ones.

Second, in the course of our analysis, it also became clear that there were a set of system elements or functions that either were not explicitly represented in our system model or were common elements of a number of different parts of the system. An example of the former is communications systems: No component in our model explicitly captures communications technologies, but their usage comes into play in a variety of ways throughout the model. An example of the latter is the set of failures that produce a shortage of resources necessary for performing a specific function in the model. Though the details of resource shortages are specific to each response task (e.g., for victim treatment, the relevant shortage may be of trained hazmat responders or of medical supplies), the classes of failures that produce such shortages can be represented in a common way, which simplified presentation and examination of the failure trees as a whole. Though we could have treated these elements differently by constructing the response model in a different way,[4] we elected to keep them as separate elements in the failure mode analysis. We labeled these trees "general" or "generic" failure trees, since they captured general response functions or failure modes. In considering the results of the analysis, these should be viewed less as stand-alone failure trees than as modules that appear as adjuncts to the failure trees directly representative of pieces of the response model.

In the remainder of this chapter, we discuss our failure mode identification for the chlorine response operation in summary, present some example failure diagrams

[4] For example, by having an explicit communications block in the diagram that as a common node between many different functional boxes in the current model. We believed that this would hurt the understandability of the model enough that it justified treating these elements separately.

in more detail,[5] and conclude with some overarching observations regarding chlorine response based on the results.

## Overview of Our Chlorine Response Failure Trees

Our examination of the chlorine response operation was structured around the main functions included in the response model. This meant that we produced individual failure trees for different components of the model, interconnected, as described above, by the dependencies between the different elements of the response. The way we broke down the failure mode analysis is shown graphically in Figure 5.2; this figure is identical to the system model shown in Figure 4.2, except that we have added letters in black circles to designate the components for which we have created failure trees.[6] The list of failure trees we created and their correspondence to different parts of the model is shown in Table 5.1. We discuss two of these failure trees in this chapter, and present all of them in Appendix D.

As discussed above, we also identified and built an additional set of generic or general functions or failure trees (included at the bottom of Table 5.1). There are three generic functions that appear in many failure mode pages: communications,[7] transportation,[8] and decisionmaking.[9] These generic functions do not appear in the system diagram, and instead have associated failure mode trees of their own. In addition, we constructed two general failure trees that relate to other elements in the model. These are "Staging," which captures the failures that could affect staging of responders and resources, and "Resource Shortages," which aggregates all of the failures that could result in too few resources performing site-level response tasks, such as victim identification and retrieval and assisting evacuees. "Resource Shortages" is affected by multiple response functions, such as responder safety and health, dispatching, and tasking resources.

---

5 All the failure mode diagrams associated with our model are included in Appendix D.

6 As explained in Chapter Four, a larger version of the system model is included as a fold-out insert in printed copies of this document and is available for download as a PDF on the RAND website: http://www.rand.org/pubs/monographs/MG994/

7 Communications functions, and therefore communications failures, come into play in each system function that must send or receive information. This includes "Information Received," "Dispatch, Task, or Reallocate Resource Packages to Site(s)," "Request More Resources from Others," and "Allocate Available Resources to Needed Tasks."

8 Transportation is similarly relevant for the movement of resources including "Dispatch, Task, or Reallocate Resource Packages to Site(s)," "Request More Resources from Others," "Allocate Available Resources to Needed Tasks," and medical "Transport."

9 Decisionmaking is a factor in site- and system-level command activities as well as some site-level activities, such as medical treatment and transport.

**Figure 5.2**
**Thumbnail Image of the Linkage of Individual Failure Trees to the Chlorine Response Operation Model**

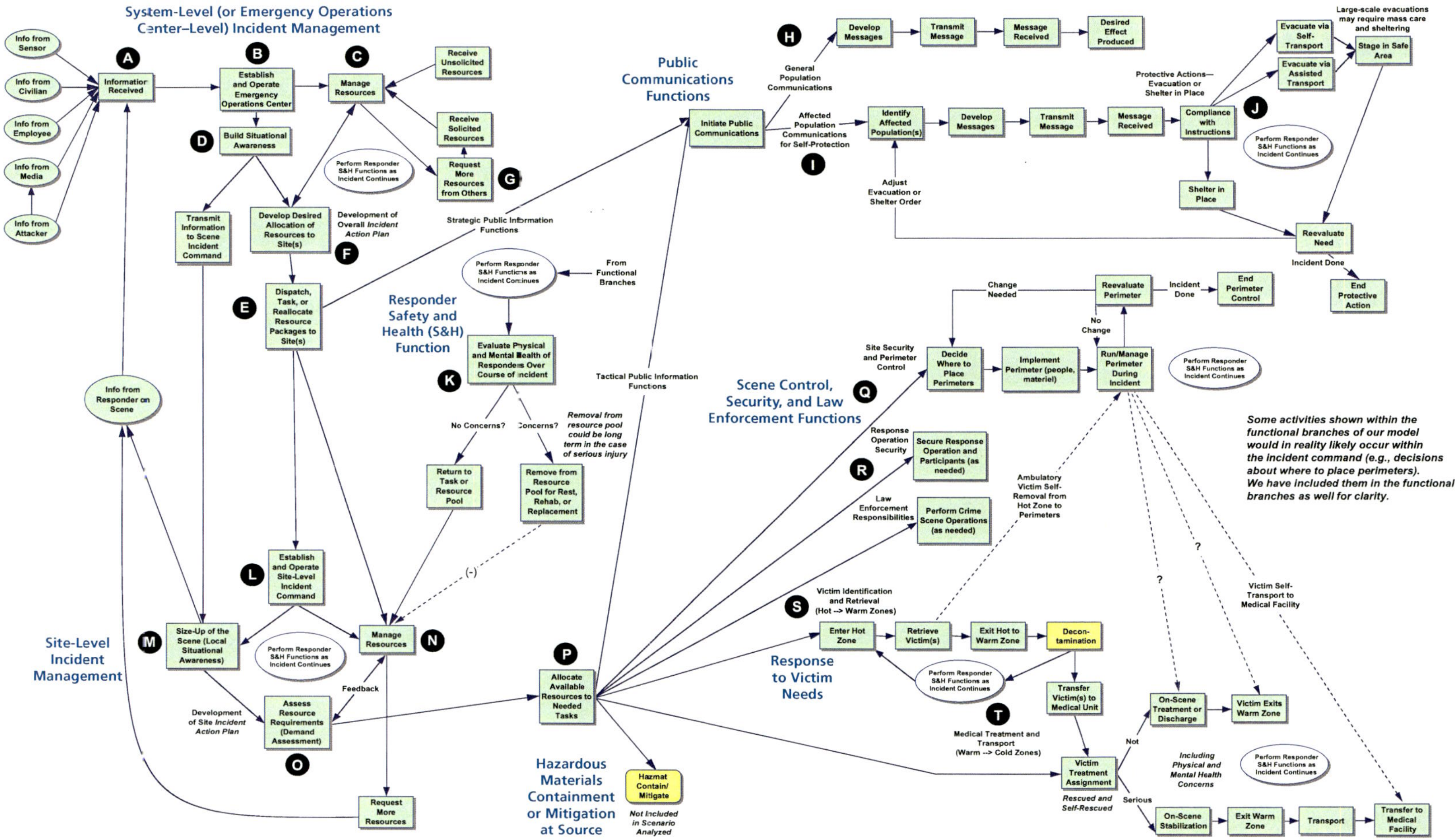


NOTES: A larger version of this figure is available as a fold-out insert included with printed copies of this document. The larger version is also available for download as a separate PDF at http://www.rand.org/pubs/monographs/MG994/.

*RAND MG994-5.2*

**Table 5.1**
**Individual Failure Trees Constructed in Chlorine Response Analysis**

| Label | Portion of Response Model | Failure Tree Title |
|---|---|---|
| A | System-Level (or EOC-Level) Incident Management | Information Received |
| B | | Establish and Operate EOC |
| C | | Manage System Resources |
| D | | Develop Picture of Incident Status |
| E | | Dispatch Specified Resources to Site(s) |
| F | | Develop Desired Allocation of Resources to Site(s) |
| G | | Request More Resources from Others |
| L | Site-Level Incident Management | Establish and Operate Site-Level Incident Command |
| M | | Size-Up Scene |
| N | | Manage Site Resources |
| O | | Assess Resource Requirements |
| P | | Task Resources According to IAP |
| H | Response Functions or Tasks | General Population Communication |
| I | | Protective Action Communication |
| J | | Evacuation and Shelter-in-Place |
| K | | Responder Safety and Health |
| Q,R | | Site Security and Perimeter |
| S | | Victim Identification and Retrieval |
| T | | Medical Treatment and Transport |
| | Generic Functions or General Failure Trees | Communications |
| | | Transportation |
| | | Decisionmaking |
| | | Staging |
| | | Resource Shortages |

## Detailed Discussion of Two Exemplary Failure Trees

As we did in the response model itself, to more specifically illustrate the construction of and content contained in the failure trees we built for a chlorine response operation, we will look in detail at two of those trees. In Chapter Four, we walked through two pieces of the response model, system-level incident command/EOC and response to victim needs. As shown in Table 5.1, those two pieces of the model have a total of nine failure trees associated with them—seven for system-level incident command (components A–G) and two for response to victim needs (components S and T). The following sections discuss the "Establish and Operate EOC" tree (component B) and "Medical Treatment and Transport" tree (component T).[10]

### Establish and Operate Emergency Operations Center

The end point failure for the establishment and operation of the EOC is framed as the command being "not functional" (shown in the green box at the lower left of Figure 5.3). This would create delays or other breakdowns in effectiveness that could affect performance in a number of other portions of the response. Working backward from that endpoint, there are four main classes of failure modes:[11]

- *EOC roles, responsibilities, or procedures not well defined*—covering problems in planning
- *EOC roles, responsibilities, or setup not executed effectively*—covering problems in implementation and technology supporting EOC operations
- *EOC not appropriately staffed*—covering staffing problems of varying types
- *EOC disrupted by incident*—covering ways that system-level management could be affected by hazards or disrupted by people during operations.

These are the main branches of the tree in Figure 5.3 (shown as yellow boxes), linking directly to the end point failure. Within each branch, more-specific causes for each of these classes of failures are broken out. In most cases, these failure modes are alternatives for one another—i.e., if any one of them occurs, the failure will occur—and so are linked by ORs in the tree. In two cases, multiple events must occur—e.g., for the EOC facility to be disrupted by the incident itself, it must be directly affected and continuity of operations plans must either be absent or fail as well. These are indicated with ANDs in the diagram.

In the system-level incident command/EOC operation failure tree, there are also three interconnections to other failure trees, shown in green boxes in the figure. They match the green box at the lower left since they are "end point failures" from other failure trees in other parts of the model. Two of these are general failure trees—one for communications (making it impossible to summon staff to the incident command) and one for transportation (making it impossible for them to report). The third box, "unauthorized people or resources allowed though the perimeter," is a linkage to the site security and perimeter failure tree, as it could result in disruption of incident management activities.

Just as these three failure trees are linked to this tree describing system-level incident command/EOC operations, the end point outcome on this tree appears in a

[10] As before, the full set of failure mode trees is included in Appendix D.

[11] Not counting the "other" category, which will be discussed in a moment.

**Figure 5.3**
**Failure Tree for "Establish and Operate Emergency Operations Center"**

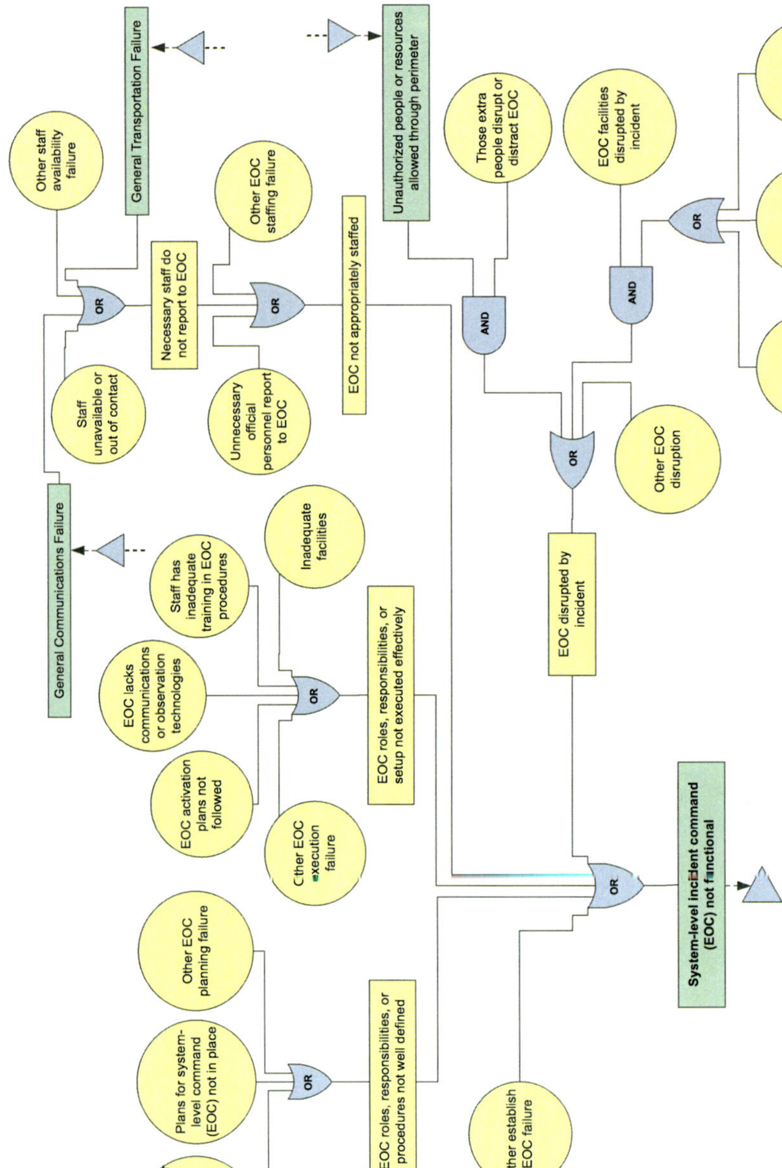


RAND *MG994-5.3*

number of other failure trees, given the role of incident command in either facilitating or directly carrying out other response functions.

Finally, in each branch, an "other" failure category is included for failures that are not called out separately. In doing an analysis of a specific response system, only failures viewed as highly unlikely would be included in the "other" category.

### Medical Treatment and Transport

In contrast to the failure tree at the incident management level, there are two end point failures on the medical treatment and transport failure tree that represent two distinct breakdowns. The first failure is serious casualties not being transported to hospitals where they can be treated. The second failure is non-serious causalities not being discharged from the scene. This second failure would potentially consume response resources (if the casualties are mistakenly viewed as serious), expose the victims to additional risk (if uninjured people or individuals who do not need intensive treatment are kept near the response scene longer than necessary), or simply increase the challenge of managing the response.

The structure of this failure tree (Figure 5.4) is more complex than the previous one, with common branches in the middle relating to both negative outcomes. The central core category failure is that injured victims do not receive appropriate on-scene medical treatment, which could result in both non-serious causalities being sent to hospital and neglecting transport of serious casualties. This could result from inappropriate treatment assignment due to number of procedural or human reasons (including a link to the general decisionmaking failure tree, in the top left of Figure 5.4), or even because victims are not successfully transferred to medical personnel (including failure to extricate them from the hazard zone—another link to another failure tree, in the top center of Figure 5.4).

The general resource shortage failure tree (top right in Figure 5.4) is linked to this tree in three ways: shortages of personnel to receive victims, shortages of personnel to treat victims, and shortage of ambulances or other equipment needed to transport victims to hospitals. In that process, the general transportation failure tree is also linked in, along with conditions at the hospitals where they are being transported that would prevent their transfer for treatment (both in the lower right of Figure 5.4).

With respect to the nonrelease of treated or uninjured individuals, potential causes of this breakdown include both the holding of the uninjured at the scene or simple failure of those manning the perimeter to release people who have been assessed and viewed as not requiring additional treatment (bottom left of figure).

**Figure 5.4**
**Failure Tree for "Medical Treatment and Transport"**

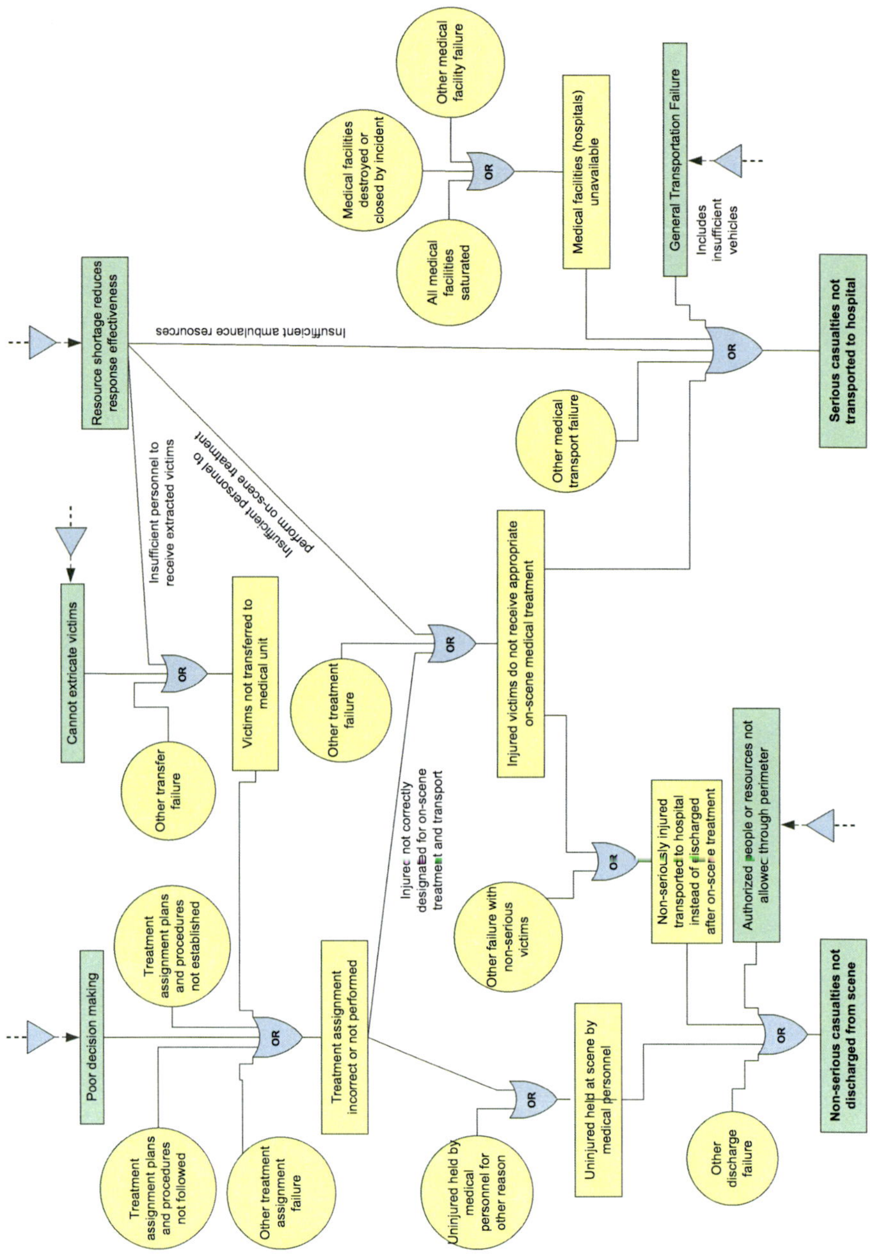


RAND *MG994-5.4*

## Discussion

Having applied the basic techniques of FMECA analysis—with some modifications—we have demonstrated that failure trees can be constructed that systematically work through the types of things that can "go wrong" in different parts of response operations and hurt performance. In Chapter Two, some initial observations about our example response system could be made based only on the identification of failure modes and which elements of the system they affected. Can a similar descriptive analysis be done on for our examination of a chlorine response? The answer is yes, though doing so is significantly more complicated than for the simple example system.

The central driver of complexity is that, unlike the linear response model used to demonstrate the analysis process in Chapter Two, the internal linkages among different parts of our chlorine response operation and different functions' failure trees discussed in this chapter are actually quite common. For readers familiar with real response operations, it is probably no surprise that, in our failure trees, the effectiveness of on-site management depends on the information and resources coming from the EOC-level incident managers, which depends in part on the information flowing up from the scene. Both depend on the functioning of communications systems in different ways, for tasks ranging from the transmission of situational awareness information to conveying response assignments to response units or staging areas. The result of this type of interaction is that the entire set of failure trees that describe the overall response operation (including all the general and specific failure trees listed in Table 5.1) contains many links between outcomes in some parts of the tree and failures in other locations in the tree. As a result, the first step in making some descriptive points about this system is to explore the linkages created by these interdependencies among different parts of the response operation. We have represented this in two ways.

First, Table 5.2 summarizes the linkages among the different parts of the overall failure tree by tracking instances where failure outcomes from one part of the tree link elsewhere. For example, looking at the second row of the table, the X's in the table show that the failure tree for the "Establish and Operate EOC" element of our model has linkages to the "Site Security and Perimeter," "General Communications," and "General Transportation" failure trees, as we discussed above. As a result, performance of the EOC- or system-level incident command is vulnerable not just to the failure modes in the part of the failure tree that affect it directly, but also to the failure modes that affect the other parts of the tree to which it is connected.

Looking at Table 5.2, we can make some general observations about failure modes in different parts of the failure tree and their interdependencies. Looking at individual rows of the table,[12] we see that the total for each row, shown in the right-most column

[12] Each reflecting the individual pieces or functions within the response model—and roughly corresponding to capabilities as described in the TCL (see Appendix B).

**Table 5.2**
**Accounting for Interconnections Among Elements of the Chlorine Response Model Failure Tree**

This portion of the failure tree: → has interconnections to these parts of the failure tree as embedded failure modes:

| | | System Level | | | | | | | Site Level | | | | | Tasks | | | | | | | General | | | | | |
|---|---|---|---|---|---|---|---|---|---|---|---|---|---|---|---|---|---|---|---|---|---|---|---|---|---|---|
| | | Information received | Establish and operate EOC | Build situational awareness | Develop resource allocation | Manage resources | Request additional resources | Dispatch resources to site(s) | Establish and operate site-level IC | Manage site resources | Size-up scene | Assess resource requirements | Task resources | Victim identification and retreival | Medical treatment and transport | Responder safety and health | Site security and perimeter | Protective action communications | Evacuation and shelter in place | General population comms | Communications | Transportation | Staging failure | Decisionmaking | Resource shortages | Row totals |
| System Level | Information received | | | | | | | | | | X | | | | | | | | | | X | | | | | 2 |
| | Establish and operate EOC | | | | | | | | | | | | | | | | X | | | | X | X | | | | 3 |
| | Build situational awareness | X | X | | | | | | | | | | | | | | | | | | | | | | | 2 |
| | Develop resource allocation | | X | X | | X | | | | | | | | | | | | | | | | | | X | | 4 |
| | Manage resources | | X | | | | X | | | | | | | | | | | | | | | | | | | 2 |
| | Request additional resources | | X | | | | | | | | | | | | | | | | | | X | X | | | | 3 |
| | Dispatch resources to site(s) | | X | | | X | | | | | | | | | | | | | | | X | X | | | | 4 |
| Site Level | Establish and operate site-level IC | | | | | | | | | | | | | | | | X | | | | | | | | X | 2 |
| | Manage site resources | | | | | | | X | X | | | | | | | | | | | | | | | | | 2 |
| | Size-up scene | | | X | | | | | | | | | | | | | | | | | X | | | X | X | 4 |
| | Assess resource requirements | | | | | | | | X | X | X | | | | | | | | | | | | | X | | 4 |
| | Task resources | | | | | | | | X | X | | | | | | | | | | | X | X | X | | | 5 |
| Tasks | Victim identification and retreival | | | | | | | | | | X | | X | | | | | | | | | | | | X | 3 |
| | Medical treatment and transport | | | | | | | | | | | | | X | | | X | | | | | X | | X | X | 5 |
| | Responder safety and health | | | | | | | | | | X | | | X | | | X | | | | X | | | X | X | 6 |
| | Site security and perimeter | | | | | | | | | | | | | | | | | | | | | | | X | X | 2 |
| | Protective action communications | | | | X | | | | | | | X | | | | | | | | | X | | | | X | 4 |
| | Evacuation and shelter in place | | | | | | | | | | | | | | | | X | X | | | | X | | | X | 4 |
| | General population comms | | | | X | | | | | | | X | | | | | | | | | X | | | | X | 4 |
| General | Communications | | | | | | | | | | | | | | | | | | | | | | | | | 0 |
| | Transportation | | | | | | | | | | | | | | | | X | | | | | | | | | 1 |
| | Staging failure | | | | X | | | | | | | | | | | | X | | | | | | | | | 2 |
| | Decisionmaking | | | | | | | | | | | | | | | X | | | | | | | | | | 1 |
| | Resource shortages | | | | X | | X | X | | | | X | X | | | X | | | | | | | | | | 6 |
| | Column totals | 1 | 5 | 2 | 4 | 2 | 2 | 2 | 3 | 2 | 4 | 3 | 2 | 2 | 0 | 2 | 7 | 1 | 0 | 0 | 9 | 6 | 1 | 6 | 9 | |

NOTES: The column totals (bottom row of the table) show the number of other failure trees in which each model element or general function appears. The higher this number, the broader the effect of problems with that element on response performance is likely to be. Row totals (the right-most column) show how many other model elements each function depends on for effective performance. The higher this number, the more vulnerable that specific element is to problems in other parts of the response.

of the table, shows the number of other parts of the failure tree connected to the identified model element or component failure tree. This provides a measure of how "vulnerable" performance in that portion of the model is to other things going wrong. In general, elements that depend on many other pieces of the model have more possible failure modes that could affect them, since the linkage to an additional part of the failure tree brings all the modes it contains with it.

The second descriptive point is that the totals that appear at the bottom of each column of the table provide a measure of the breadth of the effect of the failures in that portion of the overall failure tree. Those values are the number of interdependencies between that part of the tree and others and, therefore, the number of other places that a failure there would "propagate"—magnifying its effect on response performance.

Looking at the row totals for the failure trees corresponding to elements of the response model, though some pieces of the failure tree are interconnected more heavily to others, we see that the differences in these basic counts are not dramatic. Responder safety and health is at the top, vulnerable to failure modes in six other parts, because it is both affected by *risks* to safety created in other response activities and depends on *capabilities*, such as communications and situational awareness, to assess risk and take action to address them. Nearly half of the parts of the failure tree relating directly to response functions (i.e., excluding the failure trees for general functions listed at the bottom of the table) are linked to four or five other model component or general failure trees. None is linked to fewer than two other parts of the failure tree.

Larger differences are apparent looking at the column totals at the bottom of the table, which describe the number of other parts of the failure tree each component failure tree is connected to. Unsurprisingly, the general failure trees for communications and resource shortfalls are at the top of the list, since communications appears all over the response model and resource shortfall is a failure mode than can affect delivery of response tasks and incident management as well. Next down is site security and perimeter control, however—since breakdowns in perimeters especially are potential disruptions of management, staging, and delivery of several response tasks. Next are failures in additional general functions, transportation and decisionmaking, followed by management of the incident at the system level.

Though this type of description is a starting point for identifying the parts of the response where problems could have the broadest effects on performance, to this point we have been looking at larger pieces of the response model—rather than at the level of individual failure modes, as we did in Chapter Two. To assess how the various possible failure modes might affect system performance for each model element, we would have to do what we did for the simple example in Chapter Two—essentially, work backward from the base of the failure tree, describing the problems that could occur in that part of the system and accounting for all of the individual failure modes (including those in the failure trees of other system elements connected to it) that could produce those problems. Though doing so was a trivial exercise for our simple example case, with no intramodel links and only ten failures, the interconnections and dependencies that exist in our more realistic chlorine response model make this a difficult task at best—and in fact make it essentially impossible to provide simple totals for how many different places one failure mode "hits" response performance in the system.

To illustrate this, a more visual representation of the information in Table 5.2 is useful. Figure 5.5 shows each of the component pieces of the failure tree; arrows in the figure point *toward* a component from the other components on which it depends. For example, communications does not depend on any other functions, so no arrows point toward it. However, many arrows point outward from communications to the other parts of the tree, showing the many dependencies on functioning communications. The shapes representing the component failure trees are shown in different colors, des-

**Figure 5.5**
**Mapping the Performance Interdependencies Among Elements of the Chlorine Response Model**

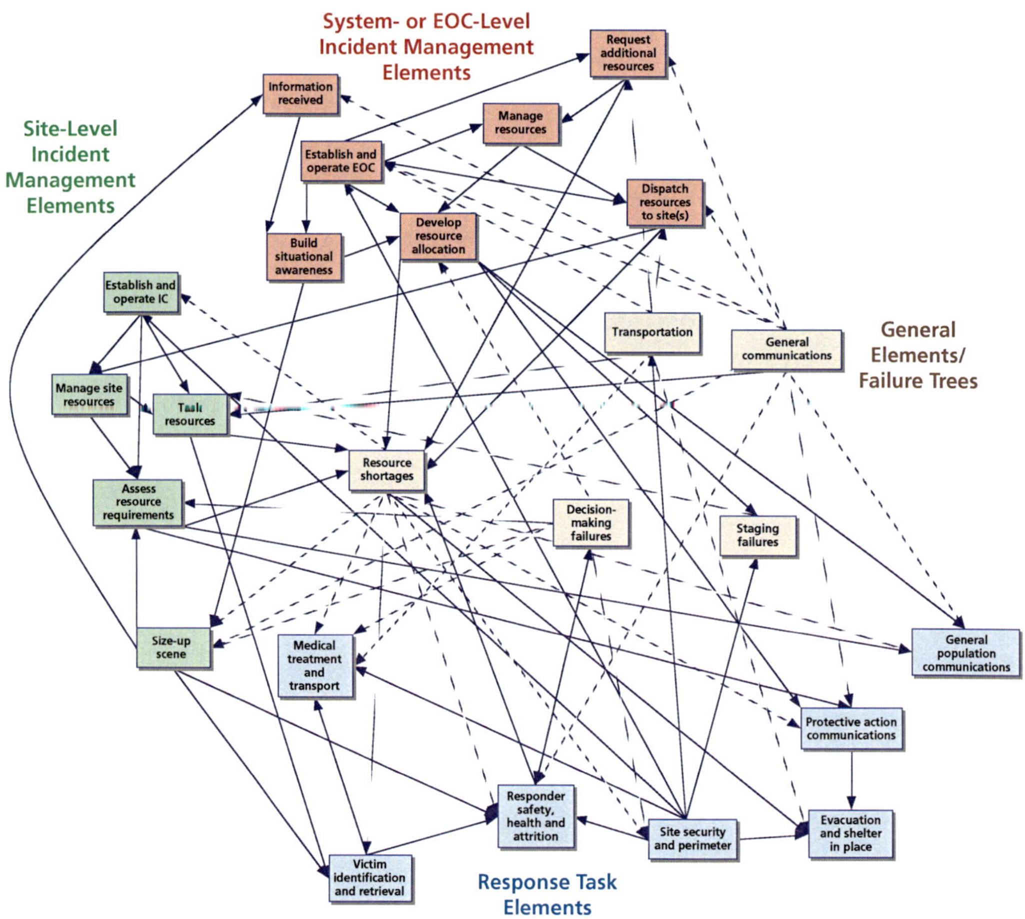


RAND *MG994-5.5*

ignating the different parts of the response model (system level, site level, response tasks) as well as the general failure trees. Arrows associated with the general failure trees are dotted to make the web of connections in the figure somewhat easier to follow.

Aside from just providing a visual representation of the complexity, picturing the interconnections in this network mapping makes it easier to go beyond what was included in Table 5.2 and, starting with one function, trace not just the others on which it immediately depends, but the functions they depend on, and so on. Doing so essentially consists of starting in a single part of the overall tree (e.g., part for a single response task) and following arrows backward to identify how failures in other parts of the tree might propogate.

The complexities associated with a more realistic response system become clear immediately. For example, starting with "Size-Up Scene" at the incident level, its functioning depends on effective decisionmaking (the link to "Decisionmaking Failures"), since the ability of response leadership to assess the scene depends on their performance. "Decisionmaking Failures" is then linked to "Responder Safety and Health," since responder fatigue (from not resting and rotating responders when needed) is one cause of decisionmaking failures. But "Responder Safety and Health" is then linked back to "Size-Up Scene," since the effective assessment of risks to responders depends on knowledge of the scene. This produces a circular dependency in the model.

Such circularities are in part due to the simplification necessary to build these types of models of complex response operations (and the fact that we are currently discussing the model at a relatively high level of aggregation). For example, like the "Medical Treatment and Transport" tree discussed previously, the "Responder Safety and Health" failure tree has two endpoints: responders injured (and therefore unavailable to continue response operations) and responder performance degraded due to fatigue (linking through to produce problems with decisionmaking and performance). This level of aggregation does not recognize the differences between those two very different (though admittedly related) outcomes and their differential effects at other points of the response model and failure tree.

However, such circularities are also *realistic*. In a complex response operation, effective allocation of resources *does* depend in part on collection of information on-scene by responders—and how much of that information is available *does* depend on whether or not responders have been tasked with collecting it (or, more realistically, whether or not the responders on scene are so occupied with delivering aid that information does not make its way back up to incident command to inform resource allocation decisions).

Given the fact that such interactions are important to capture in analysis, one potential course would be to arbitrarily set a threshold for how many "layers" to count in assessing the potential seriousness of the failure modes in different parts of the overall failure tree. The counts included in Table 5.2 represent looking one layer out—counting the interconnections between one part of the failure tree and those to which it is *imme-*

*diately* linked. The ordering of the different parts of the model/failure tree that result is the numbers at the bottom of the table. Capturing interactions one more layer out (i.e., taking a second jump across the arrow-connections in Figure 5.5) involves tabulating one additional set of linkages for each of those failure trees. Doing so increases complexity (i.e., the result is essentially Table 5.2, but expanded in an additional dimension that captures how many additional parts of the tree each of the X's in each of the individual rows is linked to), but it does have an effect. For example, because of the number of components linked to the "Resource Shortfalls" and "Decisionmaking" trees, components that they are linked to—most notably "Responder Safety and Health"—rise relative to others. Conversely, "Size Up Scene" drops relative to others since it is not linked to as many of the elements of the failure tree in the next layer of interactions.

What is the "right number" of such interactions to consider? That is an empirical question that could be explored either by looking at the effects of specific failure modes that have occurred during events (to assess directly how far their impact propagated through the response operation as a whole) or through simulation studies, such as those used in our example case. For the purposes of this discussion, we will stop here, since we address our effort to demonstrate use of empirical data as part of this type of assessment in the next chapter.

Similarly, in this discussion we have discussed the different failure trees relating to the different functions and parts in our model; we traced the links and dependencies among the whole failure trees and did not take the additional step of tabulating the individual failure modes in each of those trees and calculating the interconnections among the trees that would affect a relative ranking of the individual failure modes.[13] We have done that for several reasons, the most important of which is keeping the flow of this analysis as simple as possible at this point. For a response planner who was prioritizing targets for preparedness improvement, that next step would be essential, since approaches for addressing individual failure modes—even those within the failure trees of a single response function—are quite different. Just discussing interdependencies among sections of the overall failure tree for the response also neglects the difference between (1) modes for which a single failure is enough to impact response and (2) cases where multiple failure modes connected by an AND in the failure tree must occur simultaneously to produce an effect. In the later case, the need for mul-

[13] For some components of our response model and its associated failure trees, comparing "packages" of failure modes like this is not necessarily problematic. The more distinct the parts of the response model, the fewer similar failure modes they have in common—e.g., of the two failure trees discussed in depth in this chapter, both have failure modes associated with facilities being affected by the chlorine cloud, but those facilities are very different. One is the incident command location, and the other is the local hospital. The cases where discussing the model in this aggregated way are more of a problem is separate modes that might be corrected through common changes in preparedness or planning. For example, several places in the model and the failure trees include modes that are different in *topic* but similar in *cause* (e.g., roles not clear in the incident command, roles of operational responders for doing incident assessment not planned for) and might be addressed through similar changes in plans or training.

tiple events to happen at the same time affects the chance of the failures occurring (discussed in more detail in the next chapter) and would therefore potentially push them down a prioritized listing relative to others. This descriptive analysis—though we have drilled into the linkages between different parts of the model and resulting failure trees in some detail—is therefore not sufficient for considering priority setting. To this point, we have discussed neither the probability of failures occurring nor their consequences. We turn to those topics in the next chapter.

CHAPTER SIX

# Assessing the Probability, Effects, and Severity of Failure Modes: An Exploratory Analysis Using Response After-Action Reports

Though simply identifying failure modes and describing their impact on different parts of a response provide some insight, the picture is incomplete without the information needed to determine which failure modes are more important than others. That understanding requires estimating the probability of different failure modes occurring and assessing the consequences for response performance if they do occur.

For an actual jurisdiction evaluating the reliability of its response plan for a large chlorine release, the assessment of the probability, effects, and severity of different failure modes could potentially be done in a number of different ways. Planners in the area could assign estimates based on their past experience in the area and knowledge about what types of failures pose the greatest problems for the organizations involved. Approaches for this type of practitioner or subject-matter expert estimation could range from very informal assessment processes to much more structured activities, such as quantitative elicitation[1] or even Delphi-type[2] processes for developing consensus estimates from a group of relevant individuals. Any such assessment would essentially represent a prospective projection of future performance, structured by individual assessments of different failure modes' likelihood and consequences.

A response reliability assessment could also be retrospective, looking at performance in past incidents or in exercises and using data on specific failure modes that actually happened to support estimates of their probability and consequences. This is the process used in engineering, when there are data available from operating experience with systems or their performance in testing over time and such data can be used

[1] See, for example, discussion in Morgan and Henrion, 1990.

[2] The Delphi process was developed by RAND as a method to answer policy questions—particularly questions involving some level of subjective judgment. The technique is based on bringing together a group of experts from relevant fields and using a structured process of analytical choices to develop consensus judgments. The process includes sets of anonymous interactions focused on specific proposals or questions; the results of one round are combined, presented to the group, challenged, and then participants are given the opportunity to revise their previous answer or judgment. The process is repeated until a consensus forms.

to estimate failure rates for components or for the system as a whole. If this type of analysis was implemented as part of a preparedness assessment or corrective action program, the goal would be to gather together information on as many of a single jurisdiction, area, or even state's response operations within the past few years as possible as a starting dataset. For a jurisdiction where this type of assessment was implemented on an ongoing basis, the dataset would be updated as new incidents occurred and performance information became available, providing an up-to-date snapshot of current and likely future performance, as well as improvement or degradation in performance over time.

Performance at larger-scale incidents would, of course, be of greater relevance to understanding potential future performance in such incidents. Such a dataset could be supplemented with the results of relevant exercises, particularly if those exercises were designed to test and evaluate response capability. The failure modes observed in past incidents could then provide the basis for the various steps described in Chapter Two (from qualitative, high-medium-low–type ranking to development of quantitative estimates). Information on the consequences of those failure modes in past incidents could similarly provide the basis for estimates of the likely consequences of those failures occurring at future response operations.

To assess the utility and feasibility of conducting this type of historical analysis, we created a dataset of AARs for completed exercises and actual events. This real-world dataset on response performance provides a source with which to examine how a response reliability analysis might be done in practice. Though we will describe the sources and analytical methods used to examine our set of AARs in more detail subsequently, we found them an attractive data source for a number of reasons.

First, because of the increase in information sharing among response organizations in recent years (e.g., via DHS's Lessons Learned Information Sharing System [LLIS]), we were able to assemble a varied dataset of reasonable size easily. This included a variety of AARs relating to hazmat incidents (of direct relevance to our examination of a chlorine release), as well as other AARs for other types of incidents, allowing us to examine failure modes that were more general across response types. Second, most of the AARs provided direct data regarding actual response operations,[3] meaning that they provided information about failures that arose in the unpredictable conditions of real response activities.[4] Finally, by using existing AARs, we sought to demonstrate in

---

[3] In general, we did not examine AARs for preparedness exercises. We made two exceptions to this for two exercises that were related to hazmat, because the material included in them suggested that they had been designed such that the evaluative information provided was of comparable value to that from actual response operations.

[4] In discussing this work with response practitioners, one raised the point that "sometimes all the details, or even all the things that went wrong, aren't included in an AAR." We acknowledge that there is likely some bias in such a sample, based on what might have been omitted from post-incident reporting. In the event that a jurisdiction or area was using this approach for analysis of its own operations, this reporting bias would not be an issue, since internal information sources would be drawn on in addition to the type of data formally captured in AARs.

our proof-of-concept analysis that our response reliability assessment could be done using data that are already routinely collected by emergency response organizations. Our intent in developing this methodology was not to create an additional data requirement on response organizations to support yet another analysis; instead, the goal was to use existing data in new ways to get more insight out of those data.[5]

The following sections describe analysis of this sample of AARs, first to support assessment of the probability of different failure modes occurring and then—to a lesser extent—assessment of their consequences. We supplement the discussion by drawing parallels between our earlier theoretical reliability analyses and the chlorine response analysis.

## Exploring Failure Modes' Probability of Occurrence

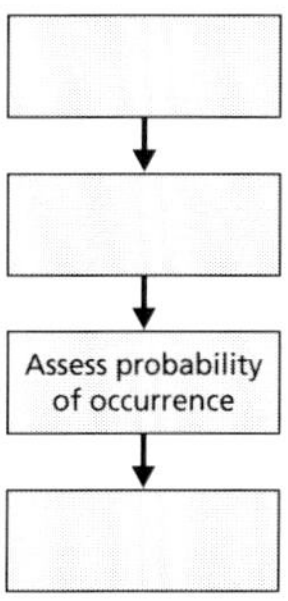


Given a dataset describing a variety of past response operations, the first step in assessing the probability of different failure modes is examining what happened in those operations, identifying what went wrong, and building descriptive statistics for which modes are more common than others.

### Description of the After-Action Report Dataset

The dataset of AARs used in this analysis includes AARs published by emergency response organizations; investigation reports by oversight boards, such as the U.S. Chemical Safety and Hazard Investigation Board (CSB), the U.S. Fire Administration, and the National Transportation Safety Board (NTSB); and articles in emergency response–related journals and magazines that discuss lessons learned for a particular incident. We gathered this convenience sample of AARs using three search methods and data sources.

First, we searched the LLIS for AARs of large-scale incident response operations and selected exercises focused on hazmat response operations. We obtained more than 100 files from LLIS, most of which fit our definition of an AAR.

Second, we downloaded all CSB investigation reports and case studies that included the words "emergency response" from the CSB website (U.S. Chemical Safety Board, no date).[6] This added another 27 documents to the pool.

Third, we searched the Internet using combinations of "after-action report" and "emergency response" as search queries to identify other promising materials that were

[5] That said, at the end of this discussion we will suggest some modifications that could be made to the information captured in these sorts of institutionalized reporting processes that would make the data more useful for this type of analysis.

[6] Accessed on April 22, 2009.

not reflected in either of the official data sources we mined. We included any AARs from federal, state, or local organizations found using this method in our sample. This third method resulted in 28 more documents that were not already in the LLIS collection.

In total, 160 documents fit our definition of an AAR. To describe the sample, we categorized the AARs by incident type (Table 6.1) and selected a subset for analysis. Because the goal of this study was to demonstrate what could be done using this data source, we selected a subset of AARs focused primarily on our scenario of interest. This meant that we analyzed 70 of the 160 AARs in our sample in detail.[7] The titles and authors or source organizations of these documents are listed in Appendix F. Because the full set of AARs was a convenience sample, conclusions drawn from the full sample would not necessarily be representative in any case, so little was lost for this demonstration by limiting the number of AARs that were fully analyzed.

The 70 analyzed AARs covered 65 different incidents. Because this study was focused on response to a chlorine release as an example case, we reviewed at least one AAR for all incidents we identified that involved chlorine or a major transportation or industrial accident, which generally involved the release of other hazardous chemicals. Of the remaining documents, we selected 21 more for review using a few criteria:

**Table 6.1**
**Characteristics of Sampled After-Action Reports**

| Incident Type | Total Collected AARs | AARs Analyzed |
|---|---|---|
| Chlorine | 13 | 9[a] |
| Industrial or transportation | 40 | 40 |
| Wildfire | 12 | 5 |
| Earthquake | 2 | 2 |
| Hurricane or tsunami[b] | 36 | 0 |
| Severe weather | 23 | 6 |
| Shooting or bombing | 6 | 1 |
| Exercise | 13 | 2 |
| Other[c] | 15 | 5 |
| Total | 160 | 70 |

[a] The four chlorine AARs not coded are additional documents describing a single incident already covered in the dataset.

[b] Fifteen of the hurricane AARs described the response to Hurricane Katrina.

[c] Includes biological incidents (4), public events (7), bridge collapses (2), building fires (1), and radiological events (1).

---

[7] One document described two different incidents and was therefore split into two separate "AARs." If the two parts of this document are counted separately, the total number of reviewed AARs is 70.

(1) since the goal was to examine failure modes that were general across response operations, the selected documents should together cover a wide range of incidents; (2) the incidents should take place in the United States;[8] and (3) the incidents should not involve events for which there was a long warning period pre-incident (for this reason, we did not include any of the many hurricane-related AARs).

Most incidents were described by only one AAR, but our sample method retrieved multiple documents for a few incidents. When we had multiple documents for a given incident, we limited the number of sources actually reviewed in an attempt to minimize the importance of any one incident in the results. We reviewed two AARs for three incidents: the Baltimore tunnel train derailment in 2007, a 2004 toxic vapor release at MFG Chemical in Dalton, Georgia, and the 2007 San Diego County firestorms. We reviewed three AARs for the 2005 train crash and chlorine release in Graniteville, South Carolina.

AARs represent an extremely heterogeneous dataset, and the quality, comprehensiveness, and information content of individual AARs can vary considerably. As a result, our attempts to balance our sample by incident type for a broader analysis were in some ways undermined by the wide variation among AARs from incident to incident. Some AARs provide extensive and in-depth analysis of incidents and what went wrong during response operations. For example, the AAR describing Seattle's response to the 2001 Nisqually earthquake is 58 pages long and includes 213 unique failures that we identified. At the opposite end, the CSB report on a 2002 explosion at First Chemical Corp. in Pascagoula, Mississippi, is 80 pages long and only describes one failure. Some AARs focus on specific elements of response. For example, one of the sources chosen describing the 2007 San Diego County firestorms turned out to be focused on serving special-needs populations during a major wildfire (we retained this AAR in the sample, given the lack of description of such issues in many of the other AARs).

CSB investigation reports were a useful source of information about hazmat incidents. However, because their focus was often on errors by the facility operators that produced the hazmat release, they did not always describe the actions of the public emergency responders involved in dealing with the incidents. Ten of the 27 CSB reports collected for this study mentioned the local emergency response but did not note failures in the emergency response process. One NTSB report also did not describe emergency response failures. It is impossible to know whether no emergency response failures were described in these documents because the responses had no failures or because the authors were not interested in the quality of the response to the incident. In part due to the variability in the quality and style of AARs, the number of

---

[8] One of the LLIS AARs was an International Atomic Energy Agency review of radiation exposure in Cochabamba, Bolivia.

failure modes identified per incident varied from over 150 in two cases to only 1–2 in many descriptions of responses to transportation or industrial accidents.[9]

### Data Analysis

We used the set of basic failure modes from all of the failure trees as a coding taxonomy for the data included the AARs.[10] To support our coding of the data, we used the textual analysis software package QDA Miner. This software provides an interface sufficient for even a complex coding taxonomy like the long list of failure modes used here,[11] which can then be selected during document review and segments of text in the document linked to one or more specific codes.[12] To illustrate the results of the coding process, Table 6.2 includes a set of examples of text elements, the failure modes they were associated with, and their sources.

Two coders performed the text analysis on different sets of AARs. Intercoder agreement was checked by comparing coding on one test case, the first 18 pages of the City of Seattle's After-Action Report for the February 28, 2001, Nisqually earthquake,[13] at the beginning of the text analysis phase. Differences in coding were discussed, and more explicit coding rules were agreed upon. Since the coders were working from different sets of documents for the rest of the analysis, no other checks for consistency were made.

Initial intercoder agreement was poor. First, when a failure in one failure tree impacted the performance of another, separate failure tree, one coder would mark the segment of text with failures from both portions of the response. The other coder marked only the primary failure, and not the implied failure relating to the other failure tree. Second, one coder used a broad definition of *emergency response organization* that included community watch groups and volunteer organizations, whereas the other had a stricter definition that included only official response organizations, such as local fire departments, FEMA, and the Red Cross. Third, one coder labeled potential failures that were issues of concern but did not actually happen in that incident, whereas the other did not. For this exploratory analysis, all of these differences were resolved

9 Details included in Appendix E.

10 As discussed previously, the initial phases of our AAR review and the development of our failure trees (Chapter Four) were developed concurrently and in an iterative manner. Based on data from the AARs, we revised the failure trees to include failure modes that had been overlooked initially. Once we felt that the failure trees had reached a level of detail and comprehensiveness that was appropriate, we froze the taxonomy of failure modes. We then returned to the analyzed AARs and reexamined them using the final taxonomy.

11 Our taxonomy included more than 250 separate failure modes across all of the failure trees.

12 Any length of text can be associated with a specific code, from a single word up to the content of an entire document.

13 We used only the first 18 pages of the test case because of time constraints and because of the usually high density of failure descriptions per page in this particular AAR.

**Table 6.2**
**Examples of Coded Failure Modes**

| Failure Tree Title | Basic Failure/Code | Segment Text | Incident | AAR Reference (see Appendix F) |
|---|---|---|---|---|
| Information Received | Sensor hardware failure | "The fire department alarm signal began sounding. However, the alarm transmitter at the CAI/Arnel facility malfunctioned before the signal code was completed." | CAI, Inc., and Arnel Company, Inc., confined vapor cloud explosion | CSB Report 2007-03-I-MA |
| Establish and Operate EOC | Staff unavailable or out of contact | "The EOC Director was on a fishing trip in the backlands of Jefferson Parish (various efforts to reach him were unsuccessful until around 8:45 a.m. when he phoned the EOC, and returned about two hours later)." | Taft, Louisiana, chemical tank explosion | Quarantelli et al., 1983 |
| Dispatch Specified Resources to Site(s) | Dispatcher error regarding resource's instructions | "The biggest criticism for the day was that a stand-by supervisor was told to respond but was not told where." | Arlington, Virginia, tanker fire | Butler, 2005 |
| Protective Action Communications | Protective action message incomplete for other reasons | "The evacuation notification process also failed to provide any specific instructions to the evacuees concerning the evacuation routes, or for obtaining updated information on the status of the evacuation." | MFG Chemical, Inc., toxic chemical vapor cloud release | CSB Report 2004-09-I-GA |
| Medical Treatment and Transport | Treatment assignment plans and procedures not followed | "Triage tags were not utilized, although they were available." | Graniteville, South Carolina, train crash | Aiken County Sheriff's Office, 2005 |
| Resource Shortages | Equipment pool damaged or depleted | "The chlorine aggressively corroded equipment involved in cleanup, including vehicles, tires, and air compressors; sometimes the parts had to be replaced after 12 hours." | Alberton Canyon, Montana, chlorine rail car derailment | Nordin, 2007 |

in favor of more coding rather than less, and the coders also agreed to apply multiple failure codes to a segment of text if the cause or effect of the failure was ambiguous.

Based on the lessons from this exploratory analysis, it is clear that broader or more-institutionalized efforts using this type of coding process for analysis of AARs would need an iterative training process to ensure inter-coder reliability in application of the taxonomy. In addition, it also suggested modifications to the coding taxonomy/failure mode descriptions that could make such analyses easier and improve coding consistency, including (1) increasing the specificity of some root failure modes and (2) creating failure codes that can be applied when the text describes a failure but not its root cause. There were also challenges in assessing whether failures in one area of response were mitigated in part or in total by adjustments in other functions, as well as in linking failures in general functions, such as communications to their effects in particular parts of response operations.

### Results

Existing AARs proved to be most useful for assessing the frequency—and, by inference, the potential future probability—of specific failure modes. In this section, we present the results of our coding analysis for both the full sample of 70 AARs and for the subset of the sample relating only to hazmat events, since they are most relevant to considering response to chlorine release incidents. Though the process and its results fulfilled the function intended—to demonstrate the viability of the analysis process using real-world data—its numerical results should be interpreted and applied with appropriate caution because of the heterogeneity of the AARs used, the nonrepresentative collection methods used to develop the dataset, and the discussion of variation in the coding process.

**Failure Modes Observed Across Incident Types.** In the full set of AARs in our sample, we identified and coded 1,213 instances of emergency response failures (Table 6.3). Only 60 of the coded failures were potential failures or near misses. The three codes used most often were "Resource does not have necessary equipment or training" (56 instances observed), "Staff has inadequate training in EOC procedures" (45), and "Communications hardware or software failure" (40). Most other failure modes were observed less than ten times in the 70 documents combined. Concurrent with conventional wisdom regarding the most common breakdowns during response operations, failures were most often in the "Establish and Operate EOC" failure tree (216 instances) and in "Communications" failure tree (184). The next most common failure areas involved tasking resources at the scene ("Task Resources According to IAP") and "Responder Safety and Health," both of which were observed 87 times.

Figure 6.1 shows the results of the coding effort categorized by the different component failure trees/response functions, including highlighting the specific modes that occurred most frequently. In the figure, each dot represents one of the failure modes included in the fault trees for each part of the response, with each dot's position indi-

**Table 6.3**
**Counts of Failure Modes Observed by Component Failure Tree in the Full After-Action Report Sample**

| Component Failure Tree | Observed Instances of Failure | Percentage of Observed Failure Modes |
|---|---|---|
| Establish and Operate EOC | 216 | 17.8 |
| Communications | 184 | 15.2 |
| Task Resources According to IAP | 87 | 7.2 |
| Responder Safety and Health | 87 | 7.2 |
| Establish and Operate Site Incident Command | 69 | 5.7 |
| Protective Action Communications | 64 | 5.3 |
| General Population Communications | 63 | 5.2 |
| Information Received | 53 | 4.4 |
| Manage System Resources | 47 | 3.9 |
| Develop Desired Allocation of Resources to Site(s) | 42 | 3.5 |
| Request More Resources from Others | 40 | 3.3 |
| Develop Picture of Incident Status | 38 | 3.1 |
| Evacuation and Shelter in Place | 34 | 2.8 |
| Resource Shortages | 32 | 2.6 |
| Manage Site Resources | 27 | 2.2 |
| Dispatch Specified Resources to Site(s) | 21 | 1.7 |
| Transportation | 19 | 1.6 |
| Site Security and Perimeter | 19 | 1.6 |
| Assess Resource Requirements | 17 | 1.4 |
| Medical Treatment and Transport | 15 | 1.2 |
| Size Up Scene | 14 | 1.2 |
| Victim Identification and Retrieval | 10 | 0.8 |
| Staging | 10 | 0.8 |
| Decisionmaking | 5 | 0.4 |
| Total | 1,213 | |

**Figure 6.1**
**Observed Failure Mode Frequency in Full Sample of After-Action Reports**

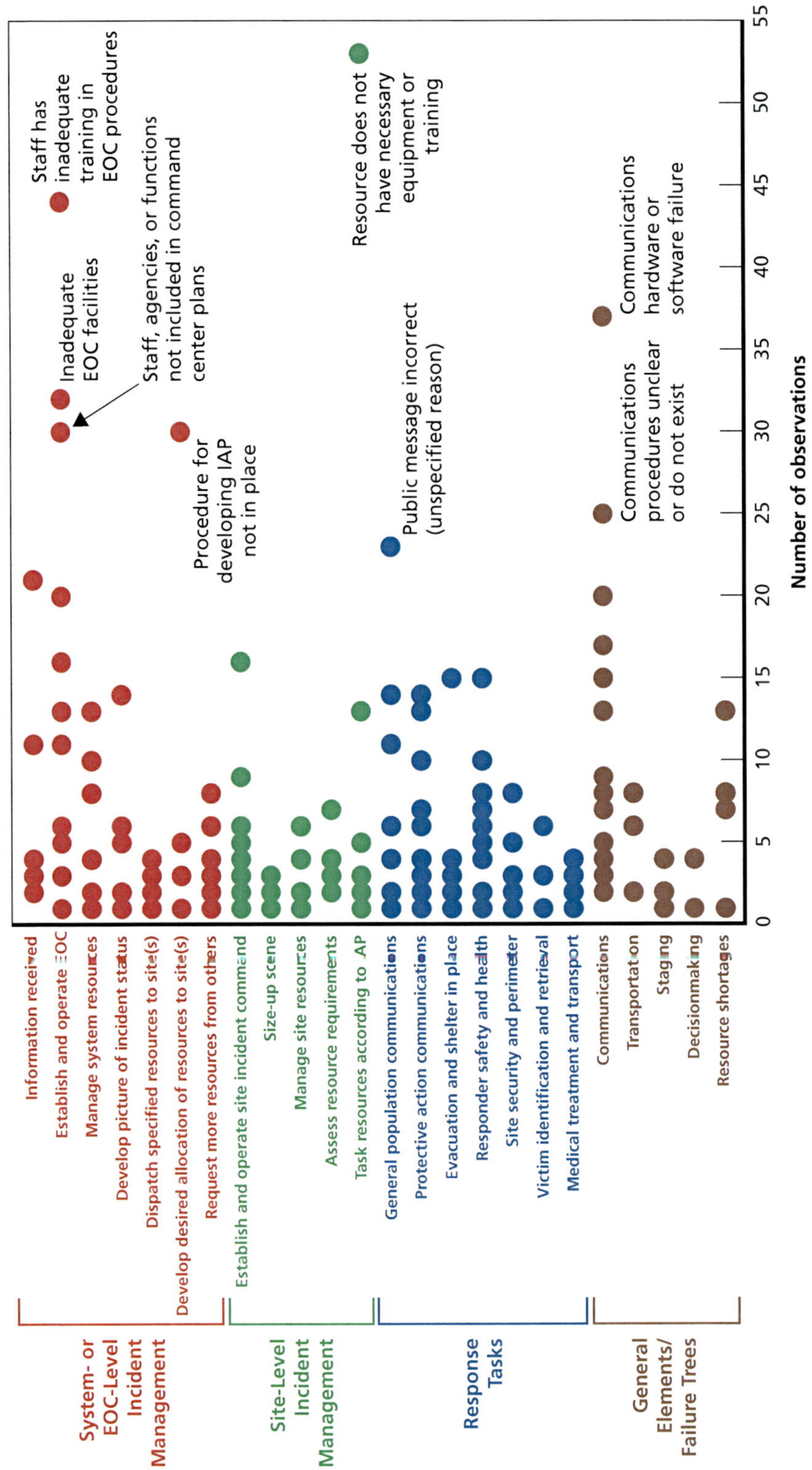


RAND MG994-6.1

cating both the part of the response to which it applies and the number of times it was observed in the set of AARs that we analyzed.

**Failure Modes Observed in Hazardous Materials Incidents.** Hazmat incidents, which included chlorine-related, biological, industrial, and transportation incidents, accounted for 553 failure codes, or a little less than half of the coded segments (Table 6.4). Thirty-seven of the 65 incidents in our sample were hazmat incidents, and hazmat incidents accounted for 41 of the 70 documents.[14] "Resource does not have necessary equipment or training" was still the most common failure mode, with 27 instances observed when only hazmat incidents were included. However, "Staff has inadequate training in EOC procedures" fell to fifth most common, with 13 instances, and "Communications hardware or software failure" fell to ninth most common, with ten instances. Instead, "Procedures for developing [system-level] IAP not in place" and "Other public/biz reporting failure" rose in relative rank, with 20 instances each.

Comparing failures observed by component-level failure tree, we find that problems in system-level or EOC incident management, with 74 instances observed, remained the most common class of failures. As might be expected given the added complexities of personal protective equipment (PPE) and evacuations in responding to chemical plumes, failures in responder safety and health jumped above communications and tasking failures to become the second most common type of observed failure.

### Discussion

Looking at the failures observed for all incidents, it is clear that the most common failures by far are those involved in establishing and operating system-level incident command and communications. Given conventional wisdom based on the experience at high-profile response operations, this is not surprising (see, for example, Donahue and Tuohy, 2006). What is perhaps more of interest is the variation in the occurrence of less common failures. Setting 20 as an arbitrary cutoff, we find that a number of failure categories are seen very infrequently in the sample. The failures that are only rarely observed include all of the key response elements directly affecting victims, though functions related to notification of the public and evacuation or shelter-in-place were more common. Looking only at hazmat incidents, we find important differences. Though system-level incident management is still prominent, it does not represent as large a share of all failure modes observed as for incidents in general. Responder safety and health concerns also represent the second most common failure mode in these incidents.

Though these basic frequency (and, by implication, future probability) values provide some insight into what failures are most likely, probabilities alone do not provide the full picture. The remainder of the chapter examines what the AARs describe about specific consequences of failures.

---

[14] The frequency with which each basic failure occurred in hazmat incidents is shown in the tables in Appendix E.

**Table 6.4**
**Counts of Failure Modes Observed by Component Failure Tree in Hazardous Materials Incident After-Action Report Sample**

| Component Failure Tree | Observed Instances of Failure | Percentage of Observed Failure Modes |
|---|---|---|
| Establish and Operate EOC | 74 | 13.4 |
| Responder Safety and Health | 61 | 11.0 |
| Task Resources According to IAP | 43 | 7.8 |
| Communications | 43 | 7.8 |
| Information Received | 42 | 7.6 |
| Establish and Operate Site Incident Command | 42 | 7.6 |
| Protective Action Communications | 33 | 6.0 |
| Develop Desired Allocation of Resources to Site(s) | 23 | 4.2 |
| Develop Picture of Incident Status | 20 | 3.6 |
| General Population Communications | 20 | 3.6 |
| Request More Resources from Others | 18 | 3.3 |
| Manage System Resources | 18 | 3.3 |
| Resource Shortages | 17 | 3.1 |
| Manage Site Resources | 17 | 3.1 |
| Transportation | 11 | 2.0 |
| Assess Resource Requirements | 11 | 2.0 |
| Site Security and Perimeter | 11 | 2.0 |
| Size-Up Scene | 10 | 1.8 |
| Victim Identification and Retrieval | 9 | 1.6 |
| Dispatch Specified Resources to Site(s) | 9 | 1.6 |
| Evacuation and Shelter-in-Place | 9 | 1.6 |
| Staging | 4 | 0.7 |
| Decisionmaking | 4 | 0.7 |
| Medical Treatment and Transport | 4 | 0.7 |
| Total | 553 | |

## Exploring Failure Modes' Effects and Severity

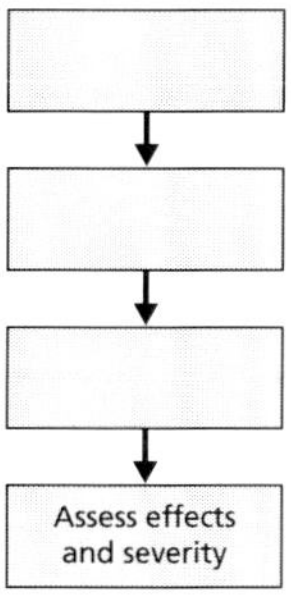


During coding, whenever a failure mode was identified, any information on its consequences was captured at the same time. In most cases, however, the AARs we reviewed frequently did not describe the consequences of individual failure modess. The clearest links between failure and consequence occurred when someone was injured. For example, the CSB report on the Little General Store, Inc., propane explosion describes one such failure:

> Guidance for emergency responders in hazardous materials emergencies recommends evacuating and evaluating the situation from a safe distance as the first task. However, the IC's [incident command's] final direction, to ensure that everyone was out of the building, came too late. Within about 30 seconds of the order, the propane ignited and the building exploded. (See Appendix F, CSB, *Investigation Report Little General Store—Propane Explosion*, 2008.)

This is an example of a failure in following site-level incident action plan procedures, which, the rest of the report explains, led to the serious injury of four employees inside the affected building.

However, in many situations, it was difficult to draw a clear link between a failure mode and any information that was provided on response performance. This was even more frequent when failure consequences were minor: In many such situations, it was difficult to determine whether there was any practical effect on the response. The report on the Taft, Lousiana, chemical tank explosion report notes this failure in perimeter controls:

> The EOC coordinator, who lives across the river, was temporarily delayed at the ferry crossing by a sheriff's deputy apparently blocking traffic ingress at that point, but she was soon allowed to proceed. (See Appendix F, Quarantelli et al., 1983.)

Since there were no major injuries or property damages to the community in this incident, it is difficult to know whether the EOC coordinator's late arrival affected activities downstream in the response or if other individuals at the EOC simply compensated for her absence.

Since our dataset from AARs does not provide the same empirical basis for discussing the severity and consequences of different failure modes with respect to our examination of a chlorine response, we are left to think through them in more categorical ways. To do so, we took two approaches. First, we examined how the interdependencies between the different elements of the response might shape the consequences of individual failures. Second, we examined individual failure modes as members of broader categories, drawing on the description of the response requirements in Chapter

Three and making arguments about how different types of failures in different parts of the response would logically affect performance of the system as a whole.

### Response Interdependencies and Failure Consequences

Before we examine how individual types of failures might result in consequences for response performance, it is worth revisiting the basic analysis at the end of the previous chapter, based only on the structure of our failure trees, and exploring how it can contribute to considering the consequences of different failures. Our counts of the number of interconnections among different parts of the larger failure tree represents one way—an extremely qualitative one—of thinking about the potential consequences of different failure modes. Failures in parts of the model that have many links to other pieces of the failure tree are more likely to have a broader effect on response operations than ones that do not.

As a result, all other factors equal, a lower-probability failure mode in a portion of the model with many connections might be of comparable concern to a higher-probability one in a less central position. For example, if we use the number of other connections to other failure trees (Table 5.2, bottom row) as a weighting factor for these observed frequencies, the relative ordering of the failure modes changes somewhat from the order in Table 6.3. After implementing such weighting, the most significant increases in perceived importance of classes of failure modes occur for resource sufficiency problems (moves up nine places in the list), site security and perimeter (moves up eight places), and transportation (moves up seven). The most significant change in the other direction is for general population communications, which drops 11 places from its position based only on the frequency data in the AARs. Though only a qualitative measure of consequences, such rough weighting does provide a way to take into account the interconnections inherent in response operations in prioritizing different observed failure modes.

### Considering Individual Failure Effects and Severity in Our Chlorine Response Analysis

For characterizing the effects and severity of failure modes, the central concern is how the occurrence of a failure will affect the outputs or outcomes that response operations are attempting to produce. In our example case discussed in Chapter Two, we explored this in a number of ways, eventually condensing our thinking into a single response reliability curve, in which the effects of failures (from ending response operations to reducing effectiveness by some percentage) were combined into a single measure of likely system performance across the full range of incident scales that could occur. Though limitations of the inferences we could draw from review of AARs about consequences of different failure modes prohibit estimation of response reliability curves that would have anything but illustrative meaning, the thought process and logic are still useful for thinking through different consequences of failures.

The starting point is to articulate, like we did for the example case, the desired final output of response activities. Unlike that case, where we worked through an example with only a single outcome (victims treated), for a realistic response to a chlorine release there would be more than one outcome. At the minimum, based on the way we have divided our response model, there would be two outcomes:

- protecting potential victims from exposure via scene perimeter control, sheltering-in-place, or evacuation[15]
- treating affected individuals fast enough to prevent their permanent injury or death.

For a chlorine response, both of these would have relatively demanding timelines: Action would have to be taken within a defined time window to be effective. In the case of preventing exposure, the time window would be defined by the period before threatened sites became affected sites (Figure 3.1). This means that delays in response action (produced either before response initiation or later, through random failure) could directly cut into outcomes, since nearer sites would be affected by the cloud and the time available to respond at farther sites would be reduced. For shelter-in-place interventions, and even more so for evacuation, there is also the potential for catastrophic failures—i.e., if people are told to shelter in areas where they are not protected, outcomes could be worse than with no response intervention. Similarly, if evacuation is begun too late, people might be brought into the open and exposed to chlorine that they might not have been otherwise.

For treating individuals, the time sensitivity is defined by the concentrations of chlorine involved and the related timelines required for treatment (Table 3.1). At nearby sites, where concentrations are likely to be higher, this compresses the window for victim retrieval and treatment. At sites farther from the source, the window could be wider if the cloud dissipates as it travels. This means that failures that delayed response activities with respect to victims would reduce effectiveness, as victims either could be injured by additional exposure or would not be treated quickly enough to address their existing injuries. Treatment delays could result from delay in action and from reductions in response capability that cut into treatment delivery rates. In structure, this treatment-over-time requirement is very similar to our example in Chapter Two, where the sole response goal was the delivery of treatment to a victim population.

As a result, looking at both of these outcomes, we find that our four classes of failure types correspond to our chlorine response operation as follows:

- *Initiation response termination.* Looking at the failure modes identified, we find that the only ones that appear likely to produce this effect would be loss of the

[15] Though we did not include it in our analysis, hazmat efforts to halt release would also contribute to this goal.

incident command through exposure, injury of a significant fraction of the response force, or disruption of operations because of significant perimeter or security failure.
- *Random response termination.* Looking across the failure modes in our model, we see that the only modes that appear likely to result in this effect would be those listed above for initiation termination failure.
- *Initiation capability reduction.* This class of failures could be produced by any initial delay that prevented starting of protective actions, such as sheltering and evacuation, or any loss of capability (e.g., through injury) that reduced the ability of the response operation to perform such tasks as evacuation assistance or victim treatment.
- *Random capability reduction.* Any delay or loss of response capability that sacrificed some portion of the available time to act or cut the rate at which people could be assisted could produce a capability-reduction failure later in the incident.

So, how do the failure modes that we have been discussing fall into these different consequence classes? Returning to the logic of the FMECA analysis, we examined each part of the response model and its associated failure tree, and identified modes that could result in the different types of consequences. The results are summarized in Table 6.5.

To construct the type of response reliability curves presented in Chapter Two for the example system, we would require the various combinations of these consequence types with their associated failure modes, along with estimates of their probability and consequences.[16] Since our dataset of AARs did not include enough information to do an empirical example of such an analysis, we do not attempt to take this final step for our chlorine response model.

#### Looking at Effects and Potential Severity in One Response Case Study

Since we could not examine the consequence side of our chlorine response model in detail, we elected to examine a single response case study for which the event, response, and issues that arose during operations were well documented. That case is the 2005 train crash and chlorine release in Graniteville, South Carolina.

On January 6, 2005, an improperly set switch caused a freight train on the main line to collide with a stationary, unoccupied train on an industrial track. The accident

[16] In addition, the effects of the transmission of failures from one portion of the model to others would also have to be defined. Given the limits of our dataset, we did not go beyond the qualitative analysis at the end of Chapter Five that involved counting the inter-linkages among different parts of the model. In a more quantitative treatment leading to building response reliability curves, a process for how a failure in incident management, for example, affects failures in the other trees would be required. The "effect" transmitted could be deterministic—e.g., a failure in incident management directly resulted in a failure of some magnitude in the other parts of the model. Alternatively, it could be more probabilistic, such that the occurrence of a failure in management increased the chance of failure in other functions but did not make those failures a certainty.

**Table 6.5**
**Potential Consequences of Failure Modes in Individual Chlorine Response Model Elements**

| Model Element/Component Failure Tree | | Failure Categories and Explanation |
|---|---|---|
| System-Level Incident Management | Information Received | ICR—all failures in tree could produce delay in starting response operations, reducing window to act.<br>RCR—failures in transmission of information from scene to system level could delay subsequent adjustment of response action. |
| | Establish and Operate EOC | IRT, RRT—branch of tree relating to system-level command being disrupted by the incident could produce termination if significant fraction of leadership was injured or facilities denied. Extended delay could be equivalent to termination.<br>ICR, RCR—all failures in tree could result in delays or poor deployment of response resources hurting protective action or victim assistance. |
| | Manage System Resources | ICR, RCR—all failure modes producing poor allocation of resources could reduce capability. |
| | Develop Picture of Incident Status | ICR, RCR—failures in building picture of incident status could delay response or produce ineffective deployment. |
| | Dispatch Specified Resources to Site(s) | ICR, RCR—failures in dispatch could delay response or produce ineffective deployment. |
| | Develop Desired Allocation of Resources to Site(s) | ICR, RCR—failures in building IAP and allocating response efforts could delay response or produce ineffective deployment. |
| | Request More Resources from Others | RCR—failures in request or delivery of reinforcing resources would cut capability. Not relevant as an initiation failure. |
| Site-Level Incident Management | Establish and Operate Site-Level Incident Command | IRT, RRT—branch of tree relating to site-level command being disrupted by the incident could produce termination if significant fraction of leadership was injured or facilities denied. Extended delay could be equivalent to termination.<br>ICR, RCR—all failures in tree could result in delays or poor deployment of response resources hurting protective action or victim assistance. |
| | Size-Up Scene | ICR, RCR—failures in situational awareness could delay response or produce ineffective deployment. |
| | Manage Site Resources | ICR, RCR—all failure modes producing poor allocation of resources could reduce capability. |
| | Assess Resource Requirements | ICR, RCR—failures in building IAP and allocating response efforts could delay response or produce ineffective deployment. |
| | Task Resources According to IAP | ICR, RCR—failures in tasking could delay response or produce ineffective deployment. |
| Response Functions or Tasks | General Population Communications | Catastrophic Exposure Prevention Failure—though most general population communications (e.g., for informational purposes) does not relate to the response outputs defined above, there is the possibility of significant failure in general population communications to result in individuals exposed to harm who would not have been otherwise. |
| | Protective Action Communications | ICR, RCR—failures in communications could delay action by responders or individuals.<br>Catastrophic Exposure Prevention Failure—as above, for general population communications, but failure in communications to either threatened or affected population creates additional exposure. |

**Table 6.5—Continued**

| Model Element/Component Failure Tree | | Failure Categories and Explanation |
|---|---|---|
| Response Functions or Tasks (continued) | Evacuation and Shelter in Place | ICR, RCR—failures delaying action or exposing threatened population to other hazards cuts effectiveness of measure to reduce exposure.<br>Catastrophic Exposure Prevention Failure—as above, for general population communications, but failure in timing or transport results in individuals exposed who would not have been otherwise. |
| | Responder Safety and Health | IRT, RRT—injury of significant numbers of responders could effectively terminate response within available response window.<br>ICR, RCR—injury of some fraction of responders or fatigue reducing effectiveness cuts performance. |
| | Site Security and Perimeter | IRT, RRT—breakdown in perimeter control or security result in disruption of key command or staging area.<br>ICR, RCT—breakdowns in perimeter management or security risks to responders produce delays or loss of resources.<br>Catastrophic Exposure Prevention Failure—as above, for general population communications, but failure in management of perimeter results in individuals entering hazardous area unnecessarily. Likely smaller potential effect that other exposure prevention failures. |
| | Victim Identification and Retrieval | IRT, RRT—sufficient delay in carrying out response activities at scene could effectively terminate response if time window is exhausted. Presumably low probability.<br>ICR, RCR—delay reduces the number of victims that can be assisted in available time. |
| | Medical Treatment and Transport | IRT, RRT—sufficient delay in carrying out response activities at scene could effectively terminate response if time window is exhausted. Presumably low probability.<br>ICR, RCR—delay reduces the number of victims that can be assisted in available time. |
| General Functions | Communications | ICR, RCR—loss of communications capability or function could reduce response effectiveness and create delay. |
| | Transportation | ICR, RCR—transportation problems could delay response. |
| | Staging | IRT, RRT—branch of tree relating to disruption by the incident could produce termination. Extended delay of access or usability could be equivalent to termination.<br>ICR, RCR—all failures in tree could result in delays of response resources. |

NOTES: When considering consequences, it is more analytically straightforward to consider the "Resource Shortage" and "Decisionmaking" failure trees as part of all the other failure trees they link with rather than separately, since it is difficult to link them directly to response outcomes in accordance with FMECA analysis, and thus we have omitted them from the table. IRT = initiation response-termination failures; RRT = random response-termination failures; ICR = initiation capability-reduction failures; RCR = random capability-reduction failures.

derailed three tank cars, each containing 90 tons of chlorine. One car was punctured and released between 40 and 60 tons of chlorine before responders patched the tank almost 24 hours into the incident. The train engineer, a truck driver at the industrial facility, six employees of a near by factory, and one person in a residence near the industrial site died from chlorine gas inhalation within minutes of the accident. Over 550 people reported to local hospitals complaining of respiratory irritation. About 5,400 people in a one-mile radius were evacuated for several days (see Appendix F, NTSB, 2005).

We reviewed three AARs for the Graniteville incident and identified 68 basic failure modes. To determine the impact of those failures on the response in terms that matched our model of response reliability, we reexamined the information available on those 68 failures to examine their consequences (both observed and potential) and timing (initiation versus random failures). The severity of the consequences was coded using the qualitative scale of response termination and serious, intermediate, minor, and negligible effect, as described in Chapter Two.

In doing so, we developed several heuristics for making consequence assignments. It was particularly difficult to determine consequences for response functions that had only an indirect relationship to lives saved, e.g., general population communications failures, such as "EOC did not have press releases prior to distribution at CP [command post]. Hard copies of press releases were not initially distributed at press conferences," from the Aiken County Sheriff's Office's AAR (see Appendix F, Aiken County Sheriff's Office, 2005). We labeled all general media coordination failures in the Graniteville case as minor, to acknowledge that they have some impact on response performance but not a direct impact on lives saved and property restored.

Failures in protective action communications, however, are potentially more serious. Still, the direct impact, in terms of the number of people who did not evacuate and were therefore injured, was not mentioned in any of the Graniteville AARs. Examples of protective action communications failures in the Graniteville incident include "Initial notification did not go out through NOAA [National Oceanic and Atmospheric Administration] Weather Radio, although it was utilized later in the day," (see Appendix F, Aiken County Sheriff's Office, 2005) and "Public unaware that unlisted phone number results in not being on 911 call list" (see Appendix F, Aiken County Sheriff's Office, 2005). Since there was no indication that these problems led directly to injuries, we also labeled all protective action failures in the Graniteville AARs as minor.

We generally labeled instances of poor coordination as minor and assumed them to be in the same category as poor logistics or personnel accountability. If the AAR stated that there was no coordination between response groups for a portion of the response, this was labeled intermediate. Units needing additional training in the incident command system were also labeled intermediate. We labeled as serious coordination failures in which an organization did not know that the EOC or site command was active or did not recognize their authority. Miscellaneous data compatibility problems such as "Standardize data collection immediately" (see Appendix F, Environmental Protection Agency, no date) were considered minor.

We considered failures to adequately monitor chlorine concentrations to be serious because missing this information risks the safety and health of responders who enter the area without proper PPE and reduces the effectiveness of hazmat cleanup activities. Words such as *chaotic* were assumed to indicate a serious failure.

Of the failures identified in the Graniteville case (Figure 6.2), we considered almost 60 percent of them to be minor or negligible in apparent consequence. We viewed only a very small minority as having potentially response-ending consequences. We categorized just over 20 percent as serious. Of those failures that were viewed as potentially response-terminating, all three were related to potential exposure of responders to hazards or resources shortages preventing decontamination activities (putting the ability to meet victim medical needs at risk).

The breakdown between initiation and random failures was heavily weighted toward events affecting performance at the beginning of response operations. We categorized approximately two-thirds of the failure modes as initiation failures.

Figure 6.3 provides a graphical summary of which elements of the response model/failure tree failures were observed in the Graniteville case study and the consequences assigned in our review. Not surprisingly, response-termination failures (or, more accurately for this review, failures viewed as having the potential of leading to response termination) were rare—the three observed fell in the victim identification and retrieval function (resource shortages associated with decontamination), the responder safety and health function (lack of hazmat monitoring in decontamination to detect hazards), and the manage site resources function (breakdowns in personnel management related to decontamination).

**Figure 6.2**
**Distribution of Failures by Assigned Consequence Level, Graniteville Response Case**

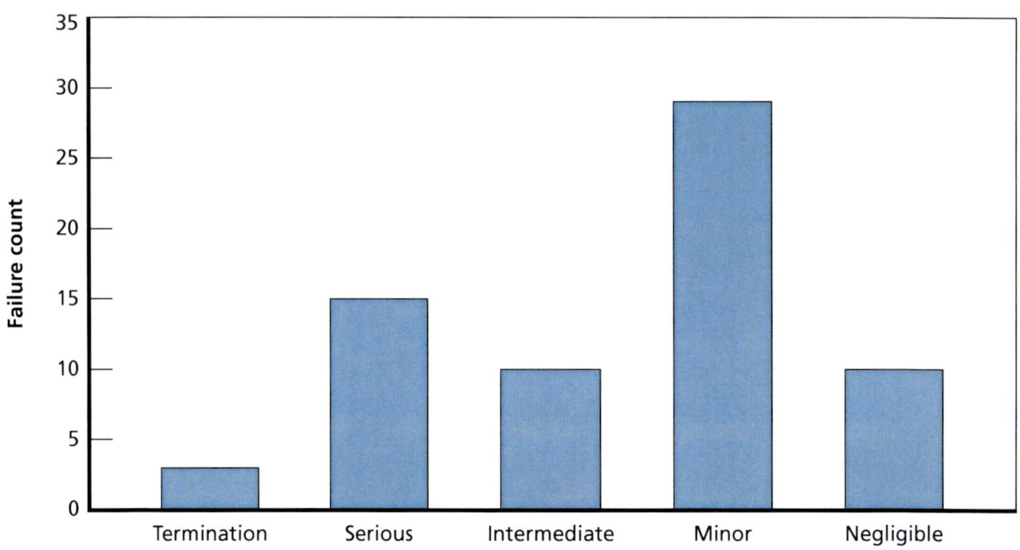


RAND *MG994-6.2*

**Figure 6.3**
**Distribution of Failures by Model Element/Component Failure Tree for Graniteville Release Response Case Study**

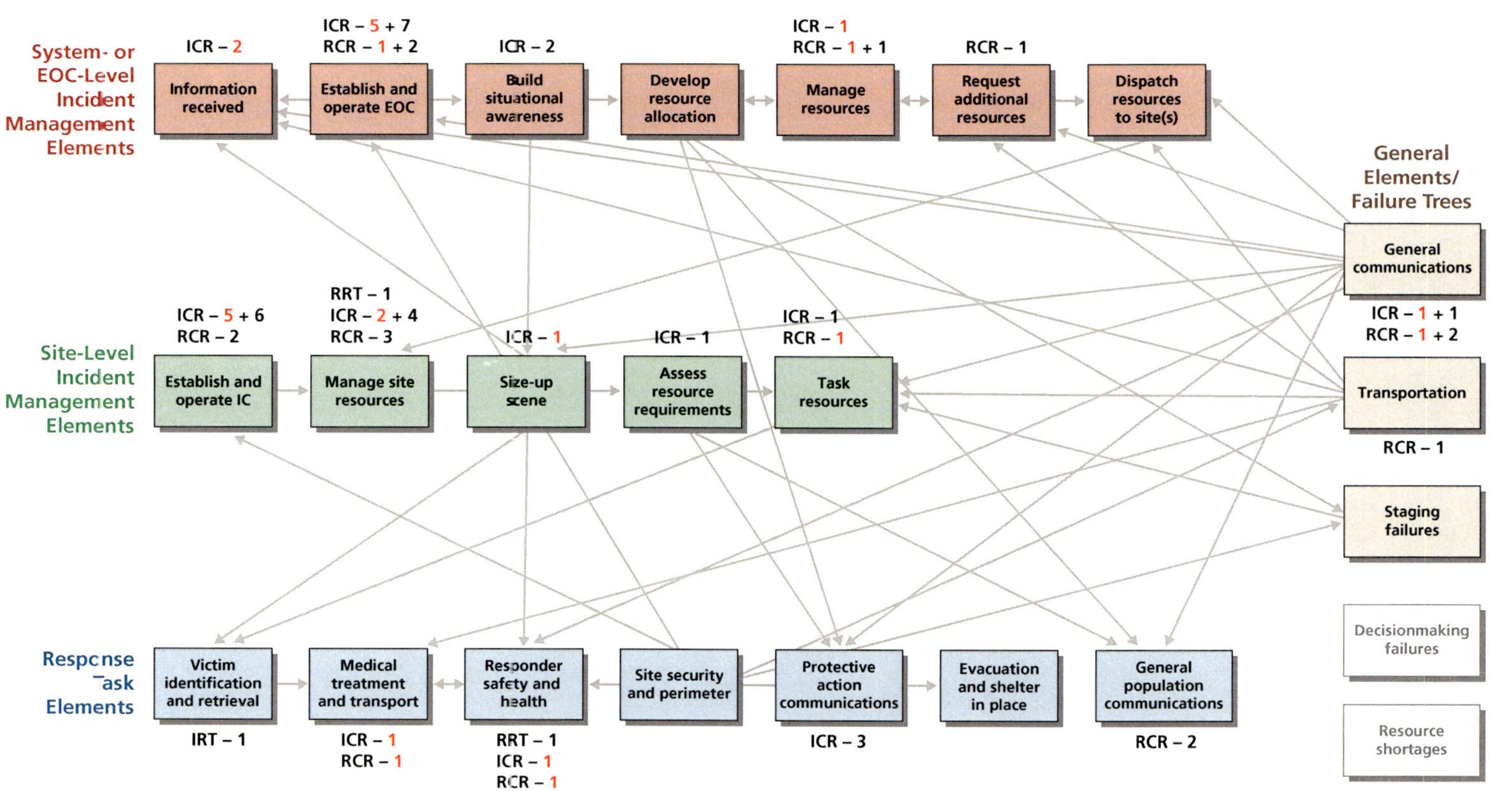


NOTES: For capability-reducing failures, red numbers indicate failures coded as serious and intermediate, and black numbers indicate failures coded as minor and negligible. IRT = initiation response-termination failures; RRT = random response-termination failures; ICR = initiation capability-reduction failures; RCR = random capability-reduction failures.

RAND *MG994-6.3*

Focusing on the intermediate and serious capability-reduction failures (designated by the red numbers in the figure), we find that those failures mostly occurred in incident management at both levels of response, though they were observed in other components as well. Though few failures were observed in the functional response task portions of our model, four of the most serious ones were in this category.

## Discussion

Using AARs as a primary source of data on response performance and failure modes observed in real response operations made it possible to test implementation of this methodology with real-world data. That effort resulted in lessons related to the method and its use, as well as some substantive observations about the failure modes observed in past response operations.

With respect to the methodology, the failure trees that were developed and their use as a coding taxonomy for primary data on response performance did provide a way to look across a very heterogeneous dataset and extract information into a common framework. This could be done in spite of the fact that individual AARs were constructed differently and included vastly different types and amounts of information. Lessons from this initial exploratory analysis—regarding the specificity of coding categories, for example—could improve future efforts at similar analyses. Though we did this analysis on AARs, both because they were available and because we wanted to demonstrate this approach using an existing dataset, a jurisdiction or area assessing its own preparedness could use internal data that potentially could be more comprehensive and have fewer limitations than our AAR dataset. The technique can also be applied using off-the-shelf qualitative analysis software packages, reducing barriers to broader use of the techniques.

In looking at the data that was available in our sample of AARs, we were able to examine frequencies of occurrence of different failure modes in past response operations—and, by implication, their likely probability at future incidents. Coupled with qualitative analyses of our model structure—specifically, how the different elements of the model depend on one another and how failures in one part could affect performance in another—such frequency data can help to prioritize investments to correct different preparedness problems. In considering failure consequences, information was scarce in most AARs, so our analysis in large part had to return to thinking systematically through the model and walking through potential consequences of failures in different parts. But even such a qualitative analysis of potential consequences can contribute to planning, for example, by sifting portions of the response vulnerable to response-termination failures versus failures with less far-reaching impacts. The absence of consequence information in most AARs also suggests ways that such documents—or other post-event data collection efforts—could be strengthened to make them more useful for future analysts doing this (or other) analyses aimed at understanding and improving preparedness.

Although our analysis was done on a convenience sample of AARs for the primary purpose of validating our methodology, a few observations based on the results of the coding can be made. Though the broad classes of failures observed most commonly were not a surprise—affecting incident command and communications—as shown in Figure 6.1, the specific failure modes showed that problems in those areas can arise via a number of different mechanisms. The most common system-level incident management failures affected staff training, though the next most common were related to facilities and planning processes. In communications, technical problems were an issue, but next came problems with procedures. The most common individual failure mode—the inability to task resources because of equipment or training problems—occurred in the component failure tree that was third on the list, below the well-recognized problems in incident management and communications. Though not at the top of the list and not including any of the most common failures, the sum of the failures observed in responder safety and health put this category relatively high as well—fourth in the full set of AARs, and second when only hazmat responses were considered.

While the nature of this dataset means that any conclusions based on the analytical results should be drawn with caution, these observations could contribute to efforts to examine individual preparedness plans. Whether the problem set observed in this set of responses to varied incidents across the country is reflected in any given locality will of course depend on its specific characteristics, but these observations are suggestive of functions—and individual failure modes that can affect those functions—that may merit additional attention.

CHAPTER SEVEN

# Concluding Observations

The premise of this work was that adapting techniques from reliability engineering and risk analysis for evaluating the performance of technical systems could contribute to better ways of evaluating preparedness and anticipating the likely future performance of emergency response systems for large-scale events. We believe that premise has largely been proven out, with both the process of such analyses and their results potentially contributing to preparedness planning and evaluation in different but complementary ways.

The methods that we have described are not entirely novel, nor is the use of concepts like response reliability unknown in previous analyses of response performance. In some ways, the type of failure mode analysis we have described could be viewed as a component that a good planner should include in any emergency planning process, that is, testing and red-teaming what might go wrong with a plan as it is developed and refined. Though doing so would represent a best practice in planning and is likely institutionalized in some jurisdictions and agencies, it is almost certainly not universal. The analytical process and techniques we have described could help to enable the broader application of such approaches.

This discussion has approached the concept of response reliability from a number of directions, including use of simulation to demonstrate how the results of the analysis could aid analysis of preparedness policy to a prototype effort to extract reliability data from response AARs. To conclude, we will revisit what we believe to be the central strengths of this analytical approach and what it brings to preparedness planning and evaluation that is missing from current approaches.

The first step of the process, defining and mapping the response, takes an explicitly systems approach to how a response operation functions. In our model, we do not distinguish which responders will perform the tasks in each part of the overall system, in terms of which organizations they are a part of or which disciplines they are trained in. In some areas, a single multifunction response agency may be able to handle most incidents; in others, the response we have mapped may involve several separate organizations. By ignoring the insignia on the uniforms of individual participants in the response, this approach lays out in black and white the potential interdependencies among organizations and how seams between them could result in response failure.

In discussing our work with one response practitioner, the comment was made that "though we are supposed to be breaking stovepipes, we still do a lot of our planning within single agencies—and this captures the problems that can still create."

The second step, systematically identifying failure modes for each part of the response model, provides a structured way for doing the type of "what-if" questioning done by experienced planners, and also for capturing the results of that process so they can be explicitly included in an organization's plan and the knowledge spread among its staff. Working down to the level of individual root failure modes also makes it easier to identify solutions to identified problems, since different failure modes—even ones within the same response function—can have very different "fixes." Similarly, even just counting up failure modes and accounting for how many parts of a response they are likely to impact can help inform prioritization, with failure modes that have broad effects on performance being of particular concern.

The third and fourth steps of the process—assessing the probability, effects, and severity of the consequences of individual failure modes—get at the information needed to identify priorities more exactly and to assess the value of different preparedness interventions to fix specific failures. In our work, we drew on existing AARs from response operations as a data source for testing this part of the analysis. The AARs we examined proved to be a challenging data source, with wide variation in the quality and breadth of individual AARs. But we were nevertheless able to apply the basic analytical process we describe, and this process made it possible to extract useful data from a very heterogeneous dataset. Though we were seeking these data to inform qualitative and quantitative measures for response performance, practitioners who we interacted with suggested other uses for such datasets as well. For example, for a specific jurisdiction, data showing that failures were adding up in a specific area could be used as a way to suggest what parts of the response system might need "preventive maintenance"—refreshers in training, particular focus in near term exercises, and so on—to reduce their chances of recurrence in the future. Such applications could contribute to meeting the requirements of emergency management standards, such those produced by the Emergency Management Accreditation Program (2007) or the National Fire Protection Association (NFPA, 2007), for structured corrective action and exercise programs to improve performance over time.

Though the nature of our AAR dataset limited the breadth of the substantive conclusions that could be drawn in our proof-of-concept analysis of response to a chlorine release, it did demonstrate that existing data that response organizations already collect can provide a significant amount of the information needed for this sort of analysis. The type of analysis we describe does not require an entirely new data collection process that would burden response organizations that already have many demands on their time and resources. The kinds of data that were lacking in the AARs—most notably, information on the consequences of specific failures during response operations on performance—suggest ways that such sources of information could be improved in the future.

But stepping away from the specific AAR sample that we examined, the general analytical process we have described could be fed by other sources of information on response performance as well. Within individual jurisdictions, there is more insight into performance issues and specific problems than is likely captured in official AARs of single incidents. Similarly, these types of analyses could also be framed at a higher level, looking at national performance in larger-scale events or the national prevalence of individual reliability concerns for different types of incidents or across incident types.

To the extent that current preparedness assessment systems and remedial action management programs (e.g., FEMA, 2009b, p. ii) capture performance information from which failure mode data can be extracted, this type of analysis could contribute to current efforts to improve preparedness assessments (such as those required by the Post-Katrina Emergency Management Reform Act of 2006 [P.L. 109-295]). We believe this approach could fill in gaps in current methods for doing such assessment. Though our analysis of AARs was retrospective and historical by definition, this type of analysis could also be done in either a more real-time or even a prospective way. In the former case, a dataset on failure modes' occurrence and their consequences for a particular response organization, jurisdiction, region, or even the nation would not be viewed as a static dataset produced from one analytical effort, but one that was updated as new incidents occurred and performance in response operations assessed. Such an implementation would be consistent with FEMA's goal to "support a living reporting mechanism that will provide an up-to-date resource on the current state of preparedness" (FEMA, 2009b, p. 1) in the nation.

Though the data available to us did not support highly quantitative analysis of the chlorine response scenario, our analysis and simulation of what might be done with quantitative response reliability values demonstrate the broader potential of reliability analysis to contribute to preparedness planning and evaluation. To the extent that response reliability curves can actually be estimated for real response organizations and their operations, they could help provide policymakers and the public with a direct answer to the question, "What is the chance that things will work next time?" that most current preparedness assessment methods cannot.

More importantly, having such a measure would help to make clear how much reliability the public should expect given current investments in preparedness, the cost of increasing it, and a means to compare different possible investments to do so—from surgically fixing known failure modes to just buying more capability to put more slack in the system to respond to an unknown future. If methods for gathering the needed data or defensible ways of estimating the variables that drive the reliability of real response systems can be developed, the result would help to advance policy debate regarding preparedness and to focus on the truly key questions in this area: not just

"How much money should we spend?" but "How exactly should we spend it?"; not just "Do we need to spend more?" but "When do we know when we have invested enough?"

APPENDIX A

# Approximating Response Reliability Curves

In Chapter Two, we generated response reliability curves for our example response system using a simulation in which both potential variation in response rates and the possibility of multiple failures occurring in a single response operation were explicitly taken into account. Taking those factors into account is necessary to make the best estimates of response reliabilities and most correct shapes for response reliability curves. Though such simulations can be done using off-the-shelf software packages,[1] in an effort to make it as straightforward as possible to apply the concepts included here, we examined whether there were ways of building more approximate response reliability curves. Our goal was to provide a way to represent preparedness data this way—given the value of showing how different failure modes do, or do not, affect performance at different incident scales—in cases where it might not be practical to do so in the best way possible.

What we wanted to develop was a method for going from a list of failure modes and estimates of their probability and consequences to an approximate response reliability curve. Doing so requires two things:

1. Making the assumption that only one failure mode will occur per response operation, thus eliminating interactions between failure modes. The negative effect of this assumption on accuracy increases as the probabilities of individual failure modes become larger.
2. Not addressing the potential for random variation in response performance over time—i.e., in the example system discussed in Chapter Two, patients are *always* treated at the average treatment rate.

Making these two assumptions allows the effects of each failure mode on the response reliability curve to be determined separately, and then added together to produce the reliability curve for the system overall.

[1] We used Microsoft Excel for the simulations discussed here, though a variety of other statistical and engineering analysis packages could be used as well.

With random variation in response performance removed, the four different types of failure modes produce highly distinctive response reliability curves (Figure A.1). For each of the separate failure types, these curves can be approximated arithmetically as follows:

- **Initiation Response-Termination Failure.** The entire curve is shifted downward by the entire probability of incidence of the failure from zero to the $RC_{max}$.
- **Initiation Capability-Reduction Failure.** The response reliability curve is only affected for incidents larger than the $RC_{max}$ minus the amount by which the failure cuts response capacity. For this discussion, we will call this value $RC_{max}$failed, since it represents the maximum response capacity when the failure occurs. Above $RC_{max}$failed, an initiation capability-reduction failure shifts the response reliability curve downward by the entire probability of incidence of the failure. For example, in Figure A.1, the curve is shifted down between 1,050 and the $RC_{max}$ since the effect of the failure shown is a 30 percent reduction in capacity.
- **Random Response-Termination Failure**. The entire response reliability curve is affected, but the amount by which the failure mode reduces reliability falls off for smaller incidents. The curve is shifted downward by the full probability of incidence of the failure at the $RC_{max}$, but gradually drops to zero effect at the far left of the curve. Arithmetically, this corresponds to the curve being shifted downward by the failure probability multiplied by a ratio of the incident size and $RC_{max}$.
- **Random Capability-Reduction Failure**. Like the initiation response-termination failure, this failure mode affects only the response reliability curve for incidents larger than $RC_{max}$failed. Like the random response-termination failure, the effect of this failure mode also changes depending on the size of the incident. Both of these effects are clear in Figure A.1, which shows that the curve begins to shift downward at 1,050, with the effect gradually increasing until $RC_{max}$ (where the curve is shifted downward by the full probability of the failure occurring). Arithmetically, this corresponds to the curve shifting downward—starting at $RC_{max}$ minus the effect of the failure—by the incidence probability multiplied by the ratio

$$(\text{incident size} - RC_{max}\text{failed})/(RC_{max} - RC_{max}\text{failed}).$$

Using these arithmetic approximations to determine how much effect each failure mode will have for incidents of different sizes, we created an overall reliability graph by adding up the effects of all the individual failure modes.

For failure modes of comparatively low incidence probability, the approximate method produces response reliability curves that are close to simple simulations (i.e., ones not including random variability in response performance). Figure A.2 shows a

**Figure A.1**
**Response Reliability Curves Produced by Different Failure Types When Response Activity Is Treated Deterministically**

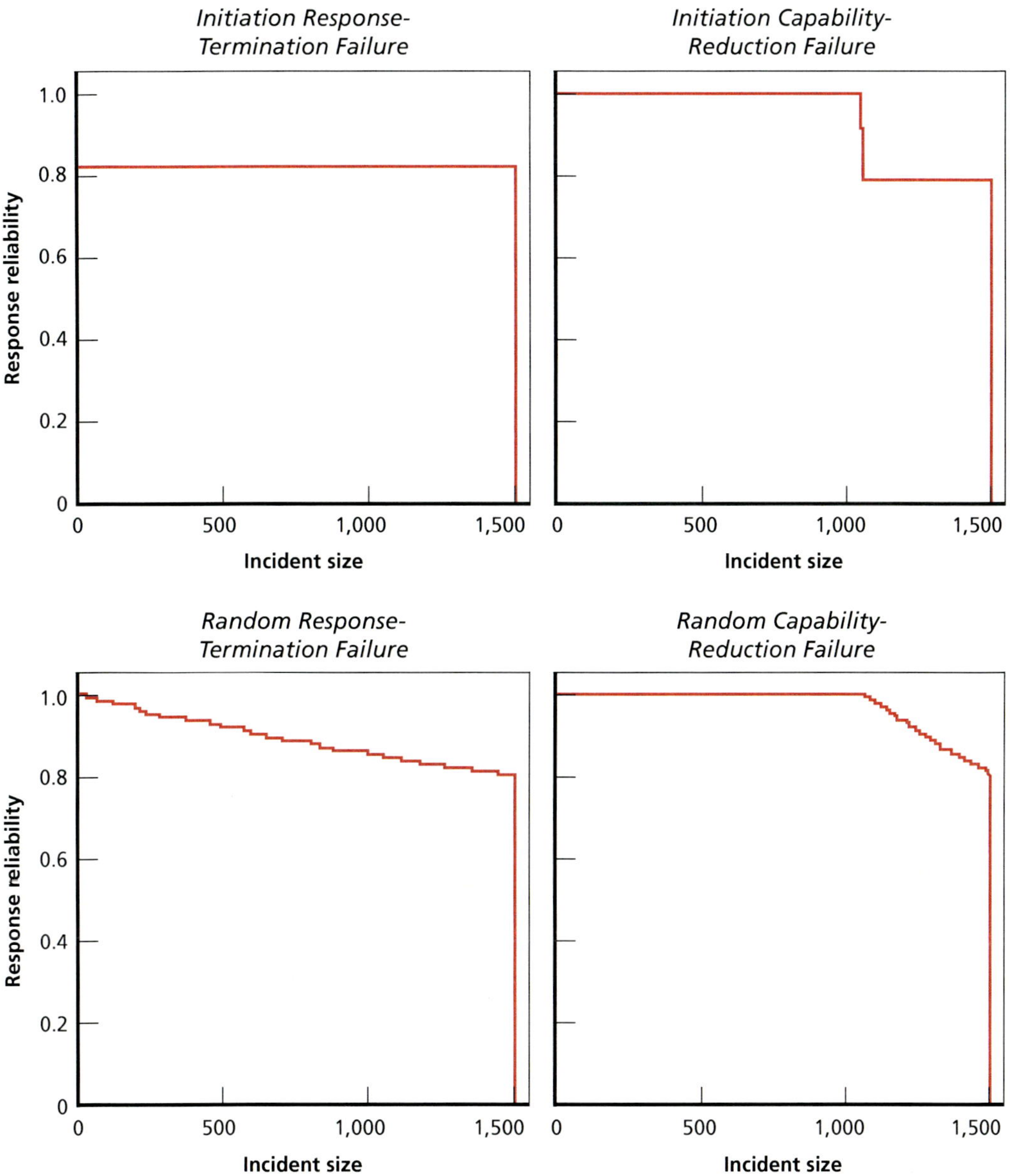


NOTES: All probabilities of incidence are set at 20 percent, and consequences of capability-reduction failures are set at 30 percent of capacity. Average treatment rate is set at 15 for all calculations.
RAND *MG994-A.1*

**Figure A.2**
**Comparison of Approximate and Simulated Response Reliability Curves: Four Modest-Probability Failure Modes**

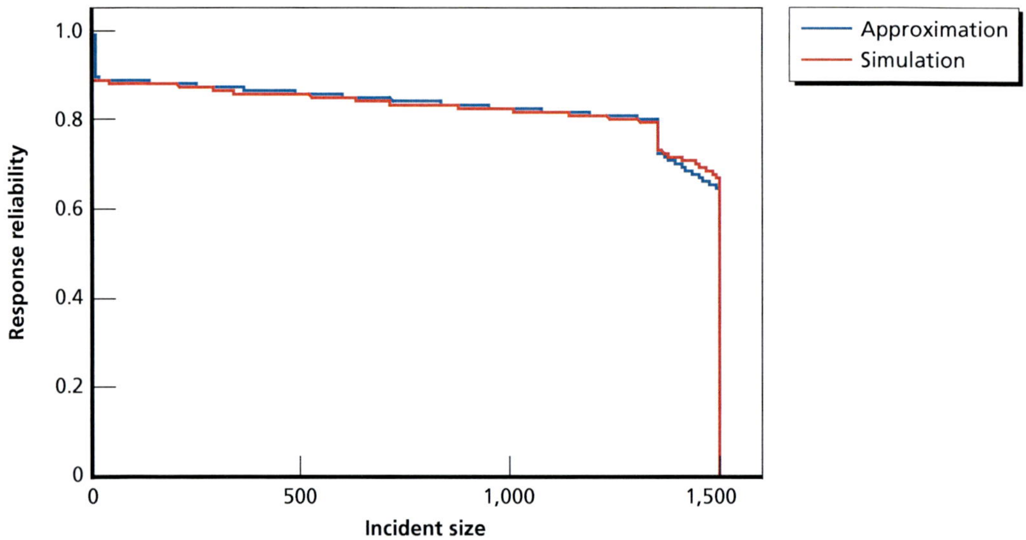


NOTES: The curves show simulation of four failure modes. Response-termination failures are set at 10 percent incidence probability and capacity-reduction failures are each set at 7.5 percent incidence probability, with a 10 percent capacity-reduction effect. Average treatment rate is set at 15.
RAND *MG994-A.2*

comparison between an approximate response reliability curve generated as described here and one generated from a probability simulation. In this example, four failure modes are included—one of each type; the probability of each type of response-termination failure is 10 percent, and the probability of the capability-reduction failures is 7.5 percent (resulting in a 10 percent reduction in capacity).

As would be expected, the approximation diverges as failure probability increases. Figure A.3 shows a case where the probabilities of incidence for both sets of failures have been doubled. Though the approximation differs significantly from the simulated value, it diverges to lower probability, meaning that it makes a conservative underestimate of the probability of future performance.

Although the results of this approximation method are conservative compared with a simple probabilistic simulation that does not take random variation in response performance into account, this is not always the case when the probabilistic simulation does include such variation. Figure A.4 compares the results of this approximate method both with those of a simple simulation (that captures the chance of multiple failures occurring at a single response) and with a simulation that both captures the chance of multiple failures and models random variation in response performance.

**Figure A.3**
**Comparison of Approximate and Simulated Response Reliability Curves: Four Higher-Probability Failure Modes**

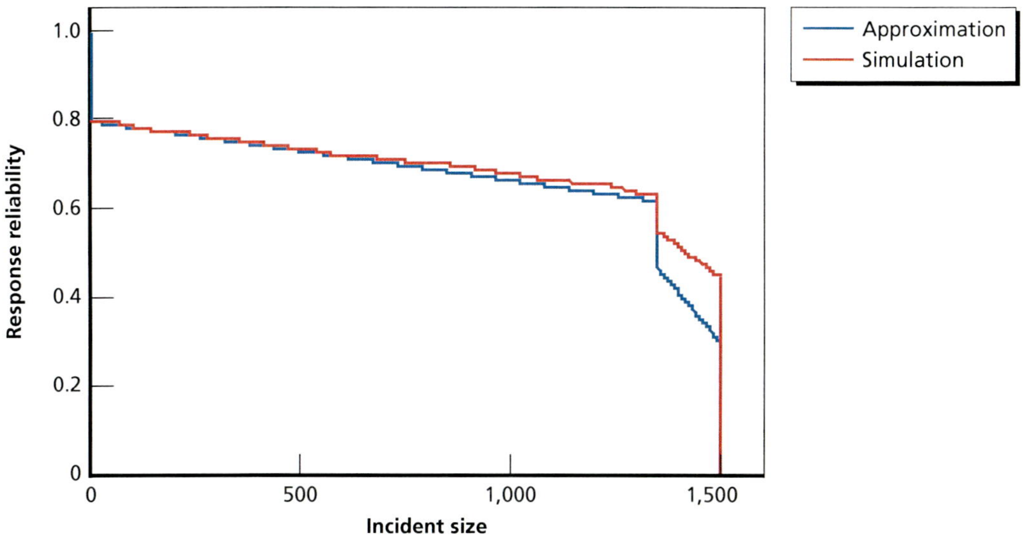


NOTES: The curves show simulation of four failure modes. Response-termination failures are set at 20 percent incidence probability, and capacity-reduction failures are each set at 15 percent incidence probability, with a 10 percent capacity-reduction effect. Average treatment rate is set at 15.
RAND *MG994-A.3*

In addition to being smoother, the simulation where response rates varied using a Poisson distribution (as in the simulations discussed in the main body of the text) falls below both of the other curves in some regions. The differences are particularly pronounced near $RC_{max}$, where the random variation included in that simulation leads to lower reliability just below and higher reliability just above that level. As a result, though this method does provide an approximate way to build response reliability curves without doing simulations, the results should not be over-interpreted.

**Figure A.4**
**Comparison of an Approximate Response Reliability Curve with Those Simulated With and Without Random Variation in Response Performance**

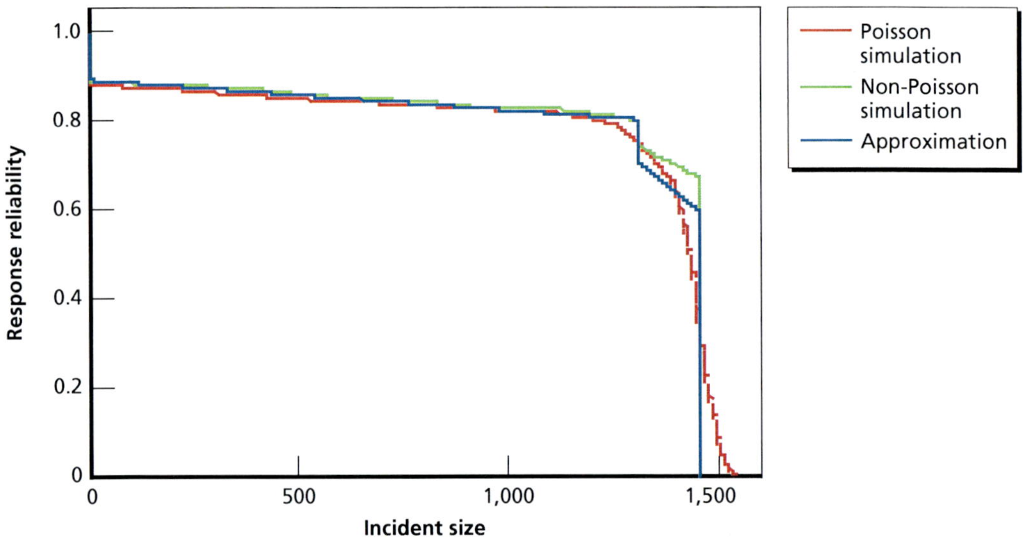


NOTES: Comparable simulations with each of four failure modes (one of each type) set at 10 percent probability of incidence. The effect of capacity-reduction failures was also set at 10 percent. Average treatment rate in all cases was set at 14.5 for comparability with the Poisson simulations, defining an $RC_{max}$ for the non-Poisson simulations of 1,450 victims.

RAND *MG994-A.4*

APPENDIX B

# Correspondence Between the Chlorine Response Model Used in This Analysis and Other Ways of Categorizing or Organizing Response Operations

In the emergency response literature, there are a number of ways both for organizing response operations themselves (e.g., the structures laid out in the incident command system in general and NIMS [DHS, 2008b] in particular for functional division of activity) and for categorizing response capabilities for planning purposes (e.g., different emergency support functions or the different capabilities laid out in the TCL [DHS, 2007b]). The structure of the response model used for this analysis differs in important respects from these other ways of considering response operations. In the interest of making the results and process described here as applicable as possible, we will briefly discuss those differences in this appendix.

The central organizing principle for emergency response operations themselves is laid out in documents describing the incident command (or management) system. At the national level, the NIMS is the overarching structure for organizing response operations—in an effort to provide a common template that makes it more seamless for response units and resources from different organizations to plug into management of a multiagency response operation. The standard structured laid out in NIMS is shown in Figure B.1.

Looking at our model of a chlorine response, the divisions between NIMS organizational locations and our functional divisions are relatively clean. Much of the activity in our functional branches would fall in NIMS's operations section. Public communications functions would be the responsibility of the public information officer on NIMS's command staff, and responder safety would be managed by NIMS's safety officer and any associated staff. The system- and site-level incident management portions of our model include both the activities of the incident command block in the NIMS chart (for the actual manager or managers of the incident) as well as the planning and logistics sections. Because of the way we scoped our model, NIMS's finance and administration section is not explicitly addressed in our model.

As part of its emergency preparedness efforts, DHS also created the TCL (DHS, 2007b), which defines a set of capabilities related to actions from prevention through

**Figure B.1**
**NIMS Organizational Structure**

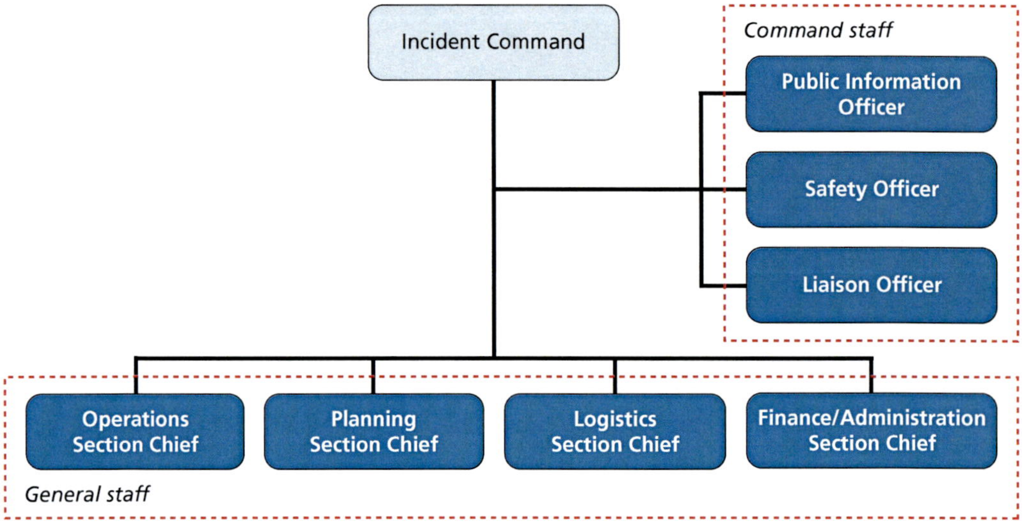


SOURCE: DHS, 2008b.
RAND *MG994-B.1*

response and recovery. Of the capabilities in DHS's "Respond" mission category, a subset of those is relevant to our example of a response to a chlorine release incident. Table B.1 includes the full listing of the capabilities and whether they are relevant to our chlorine scenario. In the table, some capabilities are listed as "possibly" relevant, since whether the capability would be needed would depend in part on how the incident was initiated. For example, an intentional chlorine release caused by a terrorist explosive would require explosive device removal as part of the response; an accidental chlorine release would not. The nature of the incident would also define the "demand" level for the other capabilities—e.g., the larger the incident, the more hazmat response capability, the more capacity to treat victims, and the more capability for effective incident management would likely be required.

As discussed in Chapter Three, because of the way that we designed our chlorine scenario and scoped our analysis, several capabilities that are relevant to response operations for chlorine releases are not included in our work. In the interest of completeness, whether we included each capability is indicated in the rightmost column of Table B.1.

Finally, just as it was possible to crosswalk the elements of our chlorine response model with the NIMS components, a similar crosswalk can be done between the structure of our model and the capabilities described in the TCL (Table B.2). Because of some limited overlap between elements of the TCL (e.g., perimeter functions for a hazmat incident could fall either entirely within the "WMD and Hazardous Materi-

**Table B.1**
**Target Capabilities Relevant to a Chlorine Release Response and Covered in Our Chlorine Scenario**

| DHS Response Capability | Relevant in Responding to a Chlorine Release? | Included in Our Scenario? |
|---|---|---|
| Emergency Operations Center Management | Yes | Yes |
| On-Site Incident Management | Yes | Yes |
| Critical Resource Logistics and Distribution | Yes | Yes |
| Volunteer Management and Donations | Possibly | Yes |
| Responder Safety and Health | Yes | Yes |
| Emergency Public Safety and Security | Yes | Yes |
| Environmental Health | Yes | Yes |
| Explosive Device Response Operations | Possibly | Partially |
| Fire Incident Response Support | Possibly | No |
| WMD and Hazardous Materials Response and Decontamination | Yes | Yes |
| Citizen Evacuation and Shelter-in-Place | Yes | Yes |
| Isolation and Quarantine | No | No |
| Search and Rescue (Land-Based) | No | No |
| Emergency Public Information and Warning | Yes | Yes |
| Emergency Triage and Pre-Hospital Treatment | Yes | Yes |
| Medical Surge | Yes | No |
| Medical Supplies Management and Distribution | Yes | No |
| Mass Prophylaxis | No | No |
| Mass Care (Sheltering, Feeding, and Related Services) | Possibly | Partially |
| Fatality Management | Yes | No |

als Response and Decontamination" capability or in the overlap of that capability and "Emergency Public Safety and Security"), some elements of our model relate to more than one capability, and a single capability may correspond to more than one piece of our model.

**Table B.2**
**Crosswalk of RAND Chlorine Response Model with DHS Target Capabilities in the "Respond" Mission Area**

| DHS Target Capability | Correspondence to RAND Response System Model |
|---|---|
| Emergency Operations Center Management | Establish and Operate Emergency Operations Center |
| On-Site Incident Management | Establish and Operate Site-Level Incident Command |
| Critical Resource Logistics and Distribution | Embedded within the "Manage Resources" components of both the system- and site-level incident commands |
| Volunteer Management and Donations | Addressed in the "Receive Solicited Resources," "Receive Unsolicited Resources," and "Manage Resources" components of the system level |
| Responder Safety and Health | "Responder Safety and Health," as well as the embedded safety and risk assessment functions within both levels of incident command and the functional branches of the response model |
| Emergency Public Safety and Security | Addressed in the three functional branches in the "Scene Control, Security, and Law Enforcement" functional branch of the model |
| Environmental Health | Addressed within the "Situational Awareness" and "Size-Up Scene" components at the system and site levels, respectively, of incident command (for environmental health hazard assessment) and implicit within the elements of "Site Security and Perimeter Control" and "Response to Victim Needs" that would be performed by hazmat trained responders equipped with analytical or monitoring equipment |
| Explosive Device Response Operations | Addressed only implicitly within the "Response Operations Security" and "Law Enforcement Responsibilities" functional branches |
| Fire Incident Response Support | Not Applicable |
| WMD and Hazardous Materials Response and Decontamination | Hot-Warm zone operations in the "Response to Victim Needs" functional branch of the model, perimeter operations in hazardous environments under the "Site Security and Perimeter Control" functional branch, elements of the "Situational Awareness/Size-Up Scene" components of both levels of incident command (for hazard assessment), and—though not applicable to our scenario—in the "Hazardous Materials Containment or Mitigation" functional branch |
| Citizen Evacuation and Shelter-in-Place | Branch of the "Public Communications" functional branch focused on both informing the public of the need for protective action and implementing that action |
| Isolation and Quarantine | Not Applicable |
| Search and Rescue (Land-Based) | Not Applicable |
| Emergency Public Information and Warning | The entirety of the "Public Communications" functional branch up to implementation of Shelter-in-Place or Evacuation measures |
| Emergency Triage and Pre-Hospital Treatment | The portion of the "Response to Victim Needs" functional branch from "Victim Treatment Assignment" to "Transfer to Medical Facility" |
| Medical Surge | Not Applicable |
| Medical Supplies Management and Distribution | Not Applicable |
| Mass Prophylaxis | Not Applicable |
| Mass Care (Sheltering, Feeding, and Related Services) | Addressed only implicitly at the end of the Evacuation branch with respect to the need for mass care in the event of an extended evacuation |
| Fatality Management | Not Applicable |

APPENDIX C

# Description of Components of the RAND Chlorine Response Model Not Covered in the Text

This appendix includes the descriptions of the parts of the system model for a chlorine response operation that were not described in the main body of the text.

## Site-Level Incident Command

The primary functions of site-level incident command are similar to those at the system or EOC level[1] but are carried out on the scene and are directly connected to delivery of response actions. The elements included in our model are diagrammed in Figure C.1 and described below:

- **Establishing and Operating Site-Level Incident Command.** At the scene, an IMS formed by the involved response organizations manages the incident. As the incident evolves, initially responding units that established the incident command are replaced with individuals or units with greater management capabilities as needed and available. As with system-level incident command, the main measure of merit for this step is the level of functionality of the incident command structure in executing the later actions required of it[2] (e.g., ensuring that such factors as friction between involved individuals and organizations does not create problems in assessing the incident or acting effectively in response).
- **Size-Up of the Scene (Building Local Situational Awareness).** The on-scene equivalent function for the system-level development of situational awareness is frequently labeled as doing a "size-up" of the incident. This process includes assessment of risk, estimation of the number of victims and their needs, and building a picture of the situation and its likely evolution over time. As for the

1 Reflected in the common structures laid out in the NIMS for management at all levels of an incident, for example.

2 As for system-level incident command, a delay in establishing incident command would be viewed as an entirely nonfunctional IMS for the purposes of this measure of merit.

**Figure C.1**
**Site-Level Incident Command Components of the Chlorine Response Operation Model**

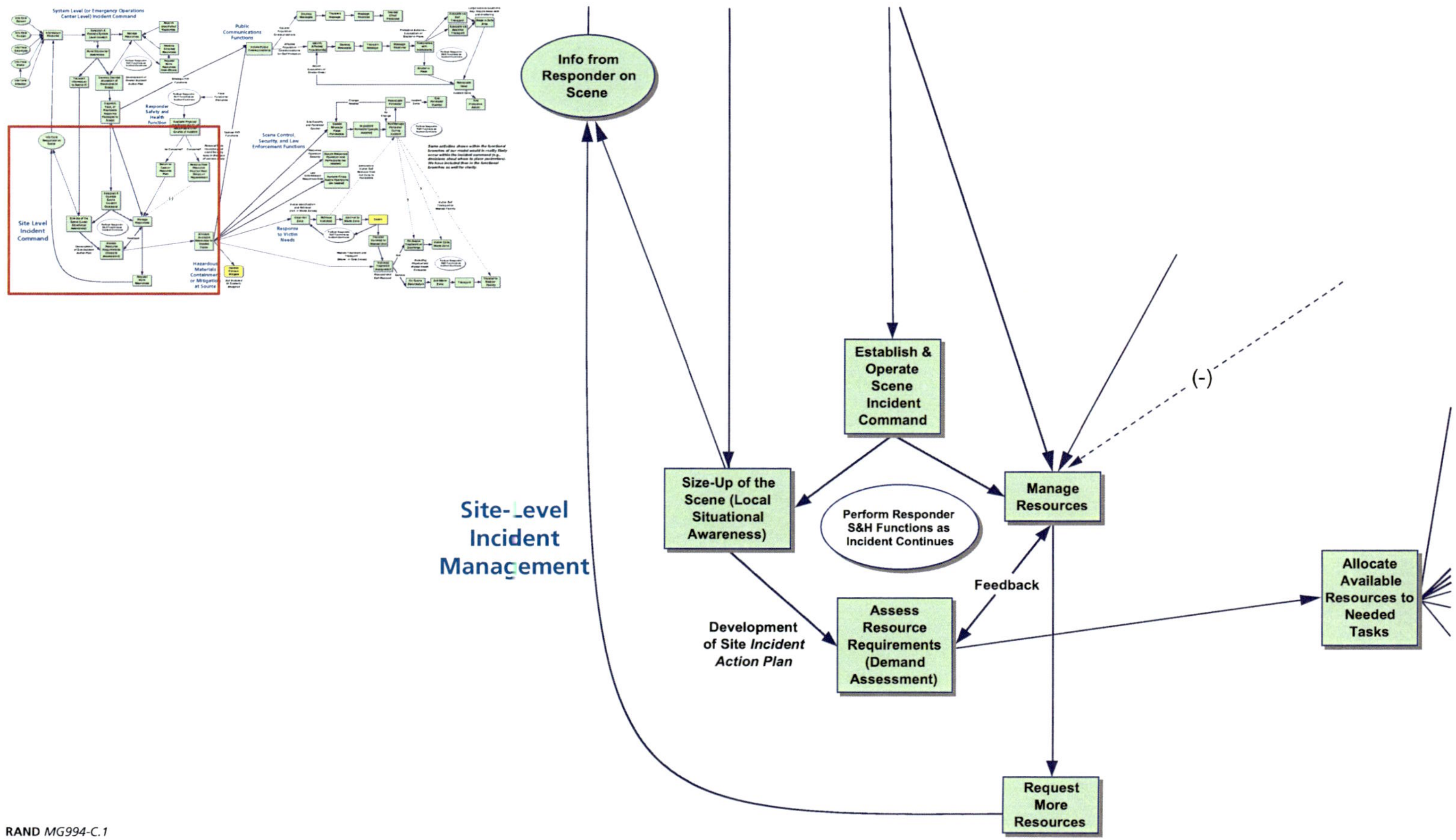


RAND *MG994-C.1*

system level, the measures of merit for this step include both the time involved and the "quality" of the size-up, since its accuracy will shape how response efforts are deployed and the needs of the incident met. In our model, we have a branch from this scene size-up, in which information is communicated back to the EOC-level incident command to inform later system-level resource allocation decisions.

- **Assessing Resource Requirements.** This function is analogous to that performed for the incident overall, but focused on a specific scene. This activity corresponds to parts of the planning functions involved in developing the incident action plan (IAP) for on-scene operations. In situations where there are excess resources, this may consist of just a matching process between resources and needs. When there is scarcity (either for limited time periods as resources are still en route to the scene or permanently because the requirements of the incident exceed the capabilities of the response system), this process will involve prioritization to get as much response outcome given available resources. As for the system level, the measures of merit are the time required and the appropriateness of the match made between the incident needs and resource requirements.
- **Managing Resources.** Just as the system-level incident command potentially had to manage resources from multiple response organizations or disciplines, the same requirement exists at the scene. It includes both informational (e.g., resource inventorying) and physical management of resources. As for the system level, the measure of merit is the "effective size" of the resource pool available to act at the scene (with maximal performance being that all resources at the scene are managed well enough that they can be rapidly and seamlessly allocated to addressing the incident requirements). When additional units arrive at the scene (in our model, dispatched from the system-level incident command), their linkage into the incident command when they arrive is a required step for them to be managed effectively as part of the available resource pool.
- **Requesting More Resources.** If the resources that are on-scene do not match the projected needs, the site-level incident command also has a route for requesting additional resources, shown in the model as a communication loop back to the system-level incident command to report that the number of deployed units is insufficient for the scene requirements. As was the case at the system level, the measure of merit for this function is whether the additional units can be requested (and arrive) rapidly enough to be of value given the incident timeline.
- **Allocating Available Resources to Tasks.** Given a match between available resources and needs, the incident command has to actually allocate and task those resources to act. The measures of merit are time required and the ability to successfully task resources to perform their assigned tasks.

## Responder Safety and Health

In our model, responder safety and health is diagrammed (in Figure C.2) as a separate function outside the incident command, although in practice the active management of responder safety—developing and implementing incident plans and actions in such a way that responders do not become casualties during their work—would be done by the incident command and the safety elements of the command staff. In our model, only the ongoing monitoring of responders in the course of their tasks is shown, with the outcome of that process being the potential for responders to be "removed" from the pool of available resources at the incident, either temporarily to maintain safety (e.g., rest and rehabilitation cycles for responders involved in victim retrieval) or as a result of breakdowns in responder protection (e.g., responders becoming injured, requiring medical attention, and being unable to continue as part of the response to the incident).

Responders being injured "in the line of duty" could occur as a result of a variety of circumstances associated with tasks performed in different parts of our model. In some situations, injuries may occur because of unpredicted (or even unpredictable) changes in incident conditions, even if reasonable measures have been taken for their protection. In other cases, injuries may occur because of breakdowns in the ways that a response is implemented (e.g., tasking responders without protective equipment to perform hazardous duties when other appropriately equipped individuals were available to do so). In our model, we include the safety management tasks involved in protecting responders—monitoring hazards, ensuring protective equipment is available and used properly, etc.—in both the incident command activities and the management of the activities in the functional branches of the model. As part of all of these functions, the obvious measure of merit is response actions being taken with as little risk to responders as possible while pursuing the response's life-safety goals.

Beyond the actions of the incident command to ensure that responders work safely, there is a safety management function that is associated with extended operations at ongoing incidents. That function is monitoring responders for exposures to hazards or things like fatigue, and pulling them from service for treatment, rest, or rehabilitation before those "transient injuries" become injuries that would require them to cease participation in the response (or even be permanently harmed). In this case, the measure of merit is the fraction of responders who should be temporarily demobilized because of fatigue, exposure, or other reasons who are actually demobilized and successfully treated before their return to duty. In our model, this later function is more explicitly diagrammed than the safety management activities designed to minimize the risk of hazardous exposure to responders during their work.

Figure C.2
Responder Safety Management Components of the Chlorine Response Operation Model

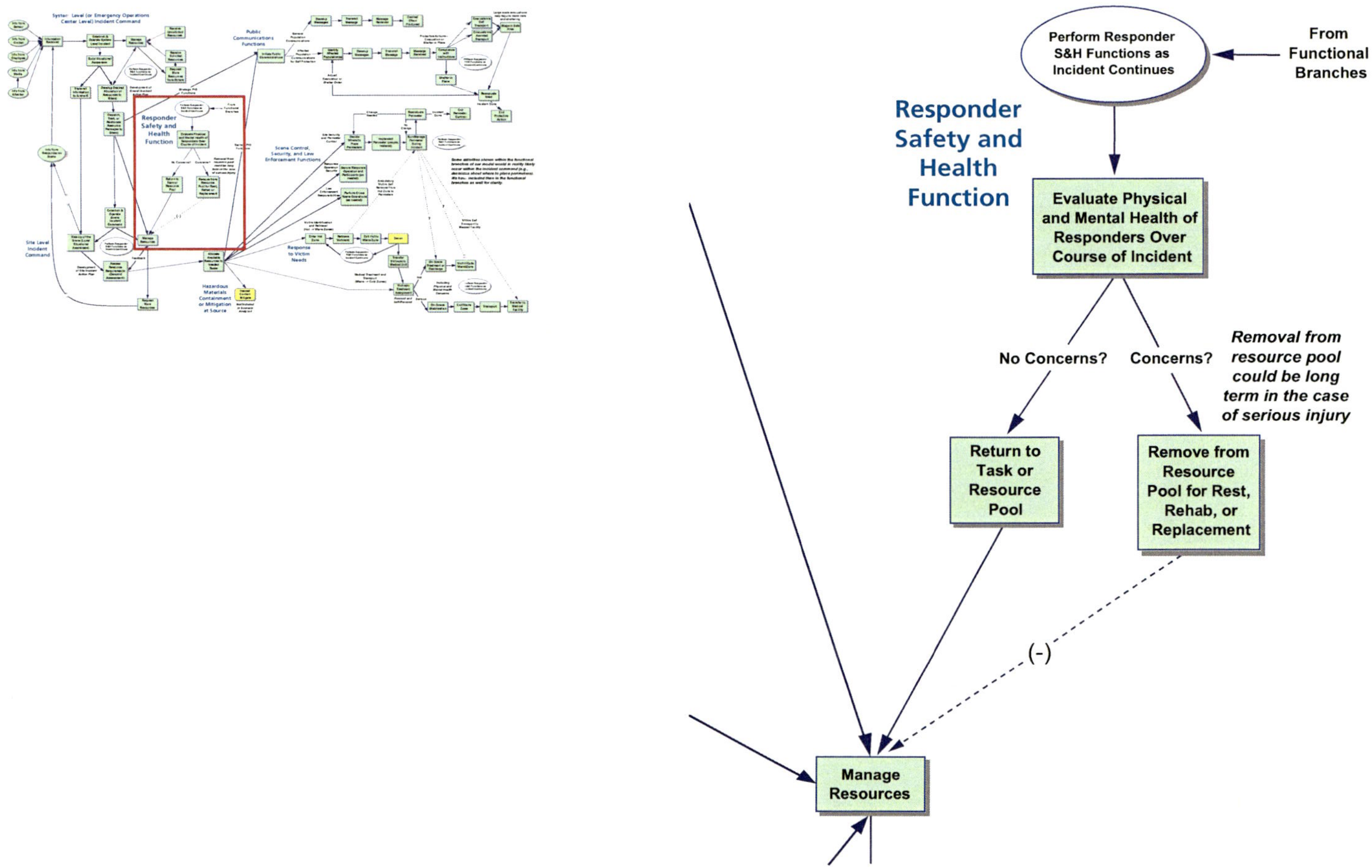


RAND MG994-C.2

## Public Communications Functions

In our model, public communications is shown as a response task that could be implemented either from the top, at the system level, or at the site level. We have included in our model both the more general public information role of informing the community and the media about the nature of the incident and progress of response (the top, shorter branch in Figure C.3, which would likely be performed at the system level) and the more specific public information function focused on giving information and direction to threatened populations regarding protective actions, such as evacuation or shelter-in-place (the bottom, longer branch, which might occur at the system level or the site level depending on the circumstances of an incident). In our model, we have a general element entitled "Initiate Public Communications" linking these two branches and serving as a placeholder for command decisionmaking (at whatever level) regarding what public information actions are required. We do not examine the more general public information function in detail since—though important—it is somewhat removed from the response actions that are focused on casualty prevention and service to affected populations.[3]

When public information interventions are focused on either evacuation or sheltering-in-place for protective purposes, an overall measure of merit for the entire branch might be either (1) what fraction of the total threatened population at the start of the incident that could be moved out of harm's way actually were moved out, or, put another way, what fraction of the total theoretical number of victims of the release were prevented from even being exposed to the hazard or (2) from the point of the evacuation or shelter-in-place decision, what fraction of the then-threatened population were successfully protected via these means.[4] In thinking about assessing the use of these interventions over the course of an incident, the second "more tactical" definition of the overall metric is more appropriate. Our individual metrics for the different model elements within this branch are framed based on that second definition:[5]

---

[3] In incidents, broader public communications efforts can have a role in minimizing the population of "worried well"—people who were not actually exposed to the hazard or were not exposed at a level that was harmful—who still seek medical care or intervention for fear that they are at risk. The challenge posed by worried well is generally viewed as affecting medical facilities most acutely—e.g., large numbers of worried well coming to emergency rooms may prevent access by those who actually need medical care. In our analysis, this issue is outside of our scope, since we have drawn the boundary of our analysis at the medical facility doors. Public communications could also have a role in warning people away from the area affected by the incident, reducing the number of potential casualties and pressure on any perimeters set up around the area of the response.

[4] Note that, for a "unitary incident" where there was a release and a single top-level decision was made about who should be evacuated or sheltered-in-place at the beginning of the response, these two definitions would collapse into one, since applying either one to that situation would be equivalent.

[5] Note that, in our discussion, each step is treated individually—i.e., there is a "time elapsed" measure of merit associated with most of the steps. In many cases, it would not be appropriate to consider the overall measure for a specific operation as the sum of all those measures. If each step was done sequentially as we have laid out (e.g., if there were only one person to do each step and he or she started on the next step only once the previous step was complete), this would be appropriate. If an emergency plan makes provisions for different tasks to be done in parallel, then the times involved will not be simply additive.

**Figure C.3**
**Public Communications Components of the Chlorine Response Operation Model**

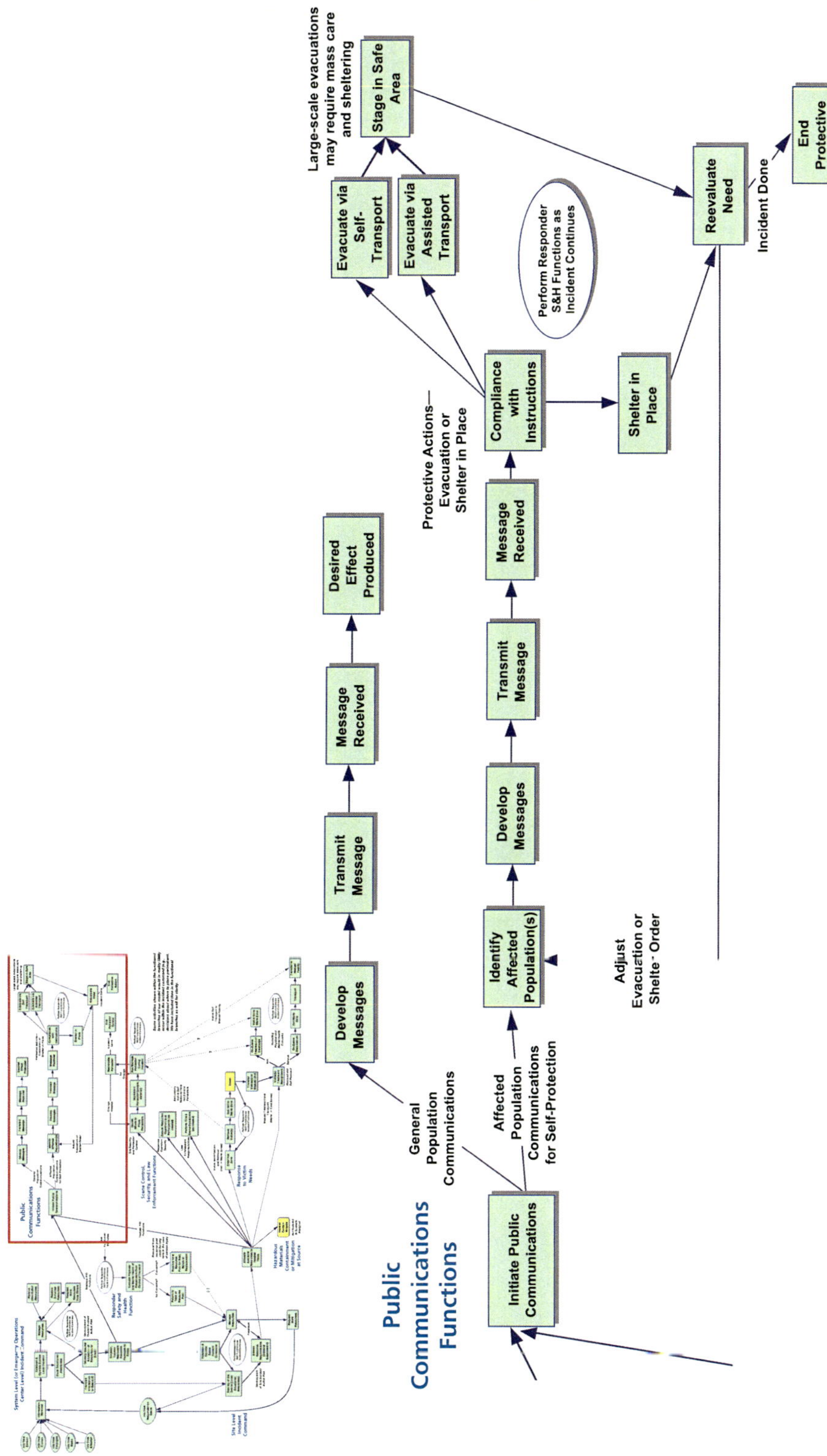


RAND *MG994-C.3*

- **Identify Affected Population**. In order for public information interventions to produce protective outcomes, they must be targeted to the populations with the potential to benefit from them. Mass evacuations where they are not needed will be disruptive, potentially costly, and could undermine the protective value of the intervention. On the other hand, very focused but mistargeted interventions (e.g., evacuating the wrong people) will have little benefit as well. As a result, for either evacuation or shelter-in-place orders, the affected population (individuals or people within an area believed to be threatened in the incident situational picture at the overall or scene level) must be identified sufficiently well that it can be contacted. For some interventions (e.g., broadcasting emergency warnings over television or radio), such identification may simply be determining what geographical areas are in the path of the cloud. For more-focused interventions (e.g., reverse 911 calls), identification could go down to the telephone exchange or even the individual level. The measures of merit for this step include the time required to perform the function (since the passage of time may limit the ability for these sorts of protective interventions) and the accuracy of the threatened population identified for communication. By accuracy, we mean that all individuals or areas that are actually threatened are included, and individuals or areas that are not threatened are not included.[6]
- **Develop Messages**. For a communications effort to be carried out, messages must be developed to communicate the information the response system needs the public to know. We assume that any response system will be able to develop such messages,[7] so the main measure of merit is the time required to develop a message that communicates the required information. For areas where the risk of chlorine release is more routine (e.g., areas around industrial facilities), message development may be done before an incident, in which case the time required for this step is essentially zero. For areas where such events are more unusual (e.g., a city where the release is an intentional terrorist event downtown), appropriate messages may have to be developed during operations.
- **Transmit Message**. For a public communications intervention to work, there must be a transmission mode (or modes) to carry the message to the targeted population. Depending on the nature of an area's planning, options for message transmission could range from broadcast by commercial media to the use of specialized alert systems (e.g., an email push alert network). As with previous steps, the time it takes for a message to be transmitted is a key measure of merit for this

[6] How this accuracy measure of merit will play out will differ from public information intervention to intervention (e.g., broad public announcement of an area-wide intervention versus focused contacts with people in a single neighborhood) and between strategic and more tactical-level information interventions.

[7] We deal with the *quality* of those messages later, in our discussion of how response operations break down—poor message quality is a failure mode, since bad messages could fail to produce the intended outcome.

step, but the outcome measures are the fraction of the affected population (i.e., the desired audience for the message) served by the selected transmission mode(s) and the ability to transmit the message successfully.

- **Message Received (by Affected Population).** Even if a message is transmitted, it may not be received and understood. For individuals to implement directions, they must actually get them and it must be clear what they are supposed to do. The measures of merit for this step include elapsed time (though the passage of time might be expected to be less of an issue in this step than in some others, and may not be within the response organizations' control) and the fraction of the target population served by the selected communications mode(s) that actually receive and understand the message.
- **Compliance with Instructions**. Because the public communication functions at issue in this branch are focused on protective actions the public can take to limit the effect of the release, compliance by the individuals who receive the directions is an important step that links the response action to a casualty-reducing outcome. As a result, the measures of merit here are the time elapsed between receipt of a message and compliance, and the fraction of the threatened population that does comply with the instructions.
- **Perform Responder Safety and Health Management Functions**. As for the system level, responder safety and health is included as a "linking function" to the specific branch of the model that involves assessing responder safety issues.

In our model, tactical public communications are focused on informing the population of the need to do one of two things: shelter-in-place or evacuate. The model elements specific to each of those actions are treated separately below.

### Shelter-in-Place Branch

- **Shelter-in-Place**. Assuming that individuals directed to shelter-in-place actually do so, whether that intervention results in reduced casualties depends on the effectiveness of the intervention for the prevailing hazard conditions. The effectiveness of sheltering in place depends on characteristics of the housing or other building stock in an area and the capability to augment structures' ability to exclude the hazard, by shutting down ventilation systems, etc. As a result, the final measure of merit for this branch of result is the theoretical maximum performance of this intervention for the incident.[8] If the concentration of chlorine involved and nature of the building stock are such that sheltering provides good protection, this measure could be very high. If not (i.e., sheltering is used for an

8 Given the characteristics of the buildings, what fraction of individuals sheltered in place within them would be *expected* to be effectively protected, for an incident of relevant characteristics.

incident for which it is not appropriate) this measure would be much lower (and casualties would be expected even among threatened populations who did successfully shelter in place).

- **Revaluate Need**. After sheltering a population in place, the incident command (whether at the system level for broad incidents or the site level for more tactical actions) will eventually have to make a decision to end sheltering or even to move from sheltering to another action, such as evacuation. This box is included in this branch as a placeholder for that task within incident command for reallocation of efforts as the incident continues and the threat environment changes over time.

### Evacuation Branch

- **Evacuate via Self-Transport.** Our model has two branches for evacuation, the first being evacuation by individuals themselves upon notification that they need to leave the area. For this step, the measure of merit is the fraction of the population that were directed to evacuate who successfully relocate out of harm's way without being exposed to the hazard. This measure includes an embedded time element, as delay in evacuation would leave people in exposed positions (e.g., in their automobiles) when the chlorine cloud reached their location, resulting in hazard exposure.
- **Evacuate via Assisted Transport.** Although many individuals will evacuate on their own, some populations will require assistance to do so. Such populations could include elderly individuals or people without transportation, but might also include facilities in the path of a chlorine release (e.g., schools) that could not evacuate the people present rapidly without assistance. The measure of merit for this step is the fraction of the population needing assistance in evacuation who are successfully evacuated without being exposed to the hazard.[9]
- **Stage in Safe Area.** The endpoint of evacuation from a hazard is shown in our model as staging in a safe area. For a small-scale evacuation, this could simply be gathering individuals in an adjacent area for the time needed for the hazard to pass. For a large-scale incident, this could involve setting up shelters and associated mass-care services to provide for larger populations moved out of the way for a more extended time for an ongoing chlorine release. Because the success of the evacuation itself is covered in the measures of merit associated with previous steps, the measure of merit for this step is the fraction of evacuated individuals who are not otherwise injured—i.e., receive injuries not associated with the chlorine release itself—during the time they are evacuated. For example, in an

---

[9] This step could be decomposed into more component elements (see, for example, notional example in Jackson, 2008). In the interests of simplifying this analysis, we have treated it as a single step. Subsequent discussion of potential failure modes for this process will suggest to the reader more specific breakdowns that could be done in a more complex systems model.

extended evacuation, if people cannot readily get food, water, or other services, it would be possible for a significant fraction to be harmed as a result. If disorder occurs at an evacuation site because of inadequate crowd control, some fraction of the evacuated individuals could be injured in the process. Such circumstances would effectively reduce the value of the evacuation as a protective intervention.

- **Reevaluate Need.** As was the case for sheltering in place, at the end of an evacuation the incident command will have a decision to make regarding whether evacuation should be adjusted (e.g., adjacent areas that were not initially evacuated now need to be as a result of changes in weather conditions) or whether the evacuation should be ended. This box in our model is a placeholder for that decisionmaking, which will take place in the incident command over the course of an evacuation and therefore does not have separate measures of merit beyond the situational awareness and decisionmaking measures already discussed above in the incident command sections.

As noted above, for both sheltering in place and evacuation, we stopped our analysis at the effectiveness of the intervention to protect against the chlorine release and did not include steps such as return of the population post-evacuation or emergence of individuals from sheltering in place. In our analysis, we have partitioned off those steps as part of recovery, since they do not have the same time-sensitive characteristics associated with them as the earlier parts of those interventions.

## Scene Control, Security, and Law Enforcement Functions

Most emergencies have a set of requirements for security and law enforcement action. We have grouped these functions into one overall branch of our model capturing scene control (i.e., creation and management of perimeters), security (e.g., controlling access to incident command sites, protecting responders doing their jobs in the event of civil disturbance or other violent threats to their safety), and law enforcement activity (in the event that the incident is a known or possible criminal act, such as a terrorist release of chlorine). These are all functions that require response personnel (in some cases, only police officers would be appropriate, whereas in other situations, other responders might be either acceptable—or in the case of hazardous environments—required). As a result, these functions might compete for personnel from other response activities. This branch of out model is shown in Figure C.4.

Perimeter control can contribute directly to reducing casualty counts by keeping additional potential victims from entering hazardous areas.[10] Site security does not

[10] Essentially, preventing otherwise nonthreatened individuals from becoming threatened or affected individuals by entering the scene.

**Figure C.4**
**Scene Control, Security, and Law Enforcement Components of the Chlorine Response Operation Model**

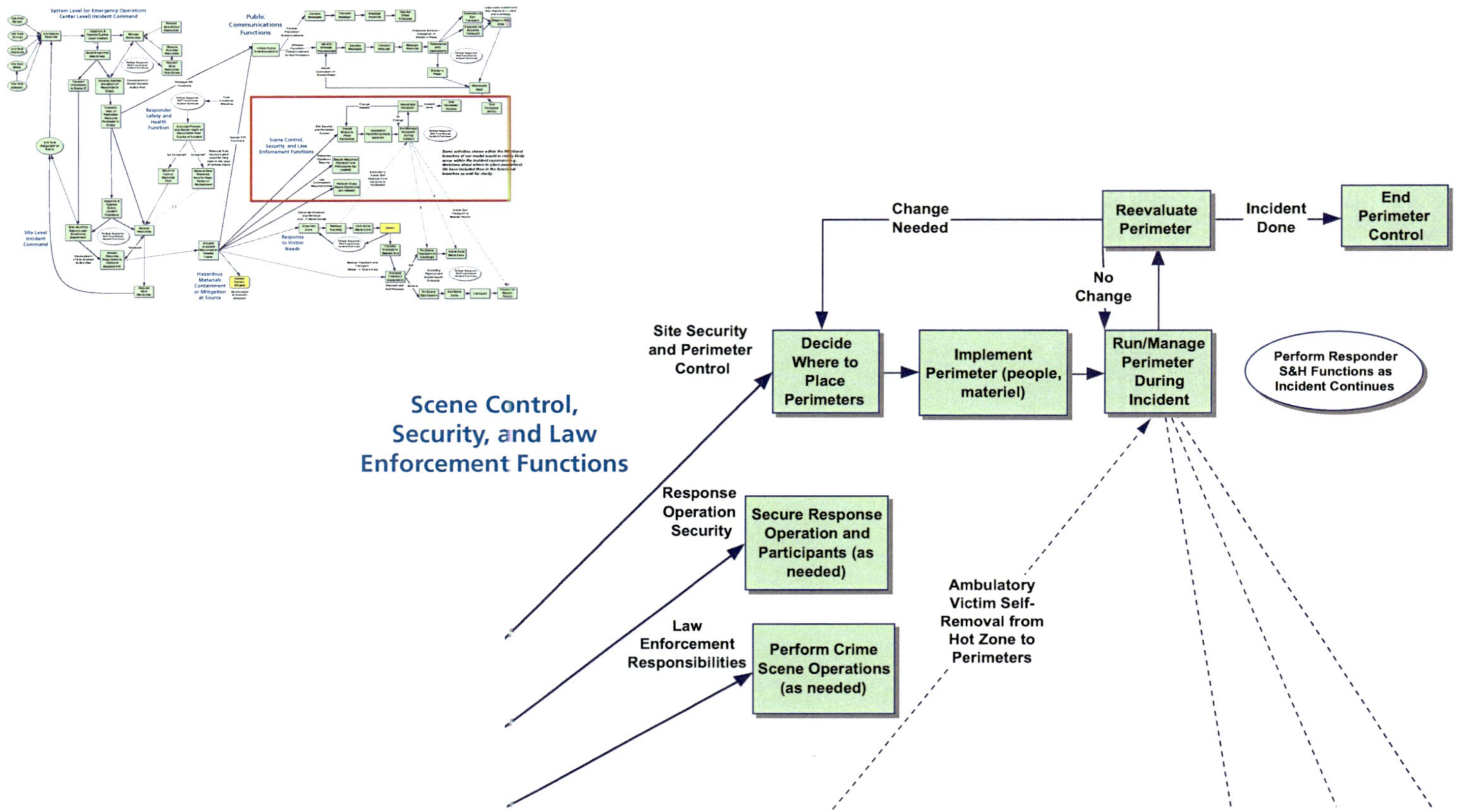


RAND *MG994-C.4*

contribute directly to meeting the needs of victims of the event, but rather addresses potential failure modes that could get in the way of responders to meet those needs (perimeter security similarly acts to address some additional failure modes).[11] Law enforcement, while necessary, is not linked to casualty reduction in the context of a single chlorine release response.[12] As a result, the measures of merit for some of these functions are related to how well they address other failure modes, rather than their own casualty-reduction effectiveness. The following sections discuss each of the components of this functional branch individually.

### Site Security and Perimeter Control Branch[13]

- **Decide Where to Place Perimeters.** As was the case with the decision step in the public information branch discussed previously, deciding where to place the perimeter(s) at an incident would be one of the resource allocation and tasking decisions made within the site incident command and is included separately in the functional branch only for clarity.
- **Implement Perimeter.** Since the goal of site perimeters is to convert an open and potentially chaotic scene into a more defined and controlled one, the key measure of merit for implementation is how fast perimeters are put into place. Within perimeters, we include divisions between hot, warm, and cold zones, as described in standard hazmat response doctrine, meaning that a chlorine release incident could involve multiple perimeters. The period between the incident occurring and perimeter implementation represents a window during which additional individuals might enter the hazard zone whose exposure could have otherwise been prevented.
- **Run/Manage Perimeter During Incident.** The key measure of performance for incident perimeters is whether they let in (and out) the people they should and deny passage to those they should not. This selective "permeability" is key to their containing the incident by keeping out individuals who should not be in a particular zone (including responders not authorized or equipped to work in the hazard environment) but not impeding response by excluding or containing

---

[11] For example, secondary or follow-on attacks on responders by the perpetrators of an intentional release.

[12] Though it could be, in the case of an intentional release where additional attacks are planned by the perpetrators.

[13] In our model, there are dashed linkages between the "response to victims' needs" branch and the perimeter control branch to address the potential for ambulatory victims to move themselves out of the hazard zone and to the perimeter without assistance. For that case, additional dotted lines go from the perimeter management box to "victim treatment assignment" (for sufficiently injured victims that need medical assistance even if they could exit the hot zone on their own), "victim exits warm zone" (for victims that are not significantly injured and can be essentially immediately discharged), and "transfer to medical facility" (since some victims—either as "worried well" or those with injuries that manifest themselves over time—will go to medical facilities on their own without transport by the response system).

people who need to cross. The measure of merit for perimeter operations is therefore the fraction of perimeter-crossing decisions that are made (letting people in or out) that are appropriate. This measure has two complementary parts: the fraction of improper actions at the perimeter that are "false positives" (people allowed to pass the perimeter who should not have been permitted) and the fraction that are "false negatives" (people denied crossing of the perimeter who should have been allowed).

- **Reevaluate, Change, or End Perimeter Control.** These are other elements included in the model for clarity that represent decisions that would be made in incident command. They include the decision that perimeters need to be adjusted (e.g., because the movement of the hazard has changed over the course of the incident) or that perimeters can be removed because the hazard is passed. These elements do not have their own measures of merit, as they are embedded in the measures for the incident command elements of the model.
- **Perform Responder Safety and Health Management Functions.** As above, responder safety and health is included as a "linking function" to the specific branch of the model that involves assessing responder safety issues.

### Response Operations Security Branch

- **Secure Response Operation and Participants (as Needed).** In some responses, there may not be a need for a dedicated response operations security effort. In others, however, threats from crime or civil disturbance could negatively affect both the effectiveness of the response and the safety of the responders involved. This function would include security activities associated with terrorist (or suspected terrorist) releases of material, such as searching for secondary explosive devices or follow-on attacks against responders. Response security operations could pull individuals away from actually meeting victim needs (e.g., if response teams must consist of more people than planned). As a result, the need to devote personnel to this function may hurt the overall ability to reduce casualties. If we accept this potential reduction in total response effectiveness, the measures of merit for this function on its own are (1) how effectively security efforts cut reductions in response effectiveness that would otherwise have occurred in their absence and (2) how effectively security reduces responder injuries that would otherwise have occurred in their absence.

### Law Enforcement Responsibilities Branch

- **Perform Crime Scene Operations (as Needed).** If an incident is (or is suspected to be) a criminal act, law enforcement officers will need to perform crime scene and other law enforcement–specific operations. Since these activities are not linked directly to the prevention of casualties in the incident itself, there are not measures of merit associated with this function in our model. Instead, we view these activities as possible competitors for resources that could otherwise have contributed to other response functions (e.g., police officers who were therefore not available to maintain the perimeter or scene security) and concerns that might limit the speed or effectiveness of other response actions because of the need to preserve evidence for later investigative purposes.

APPENDIX D

# Failure Trees for All Elements of the Response Model

This appendix presents all of the failure trees that we developed for the emergency response model. There are 23 failure tree diagrams; the diagrams that describe response or incident management activities are each associated with a subsection of the system diagram. The system diagram is presented in Figure 4.2 in Chapter Four; a larger version is included as a fold-out insert in printed copies of this document and is available for download as a PDF on the RAND website.[1] Figure 5.2, the fold-out insert, and Table 5.1 show how each failure tree relates to an area of the system diagram. The letters used in Figure 5.2, the fold-out, and Table 5.1 are shown in black circles next to the title of each of the corresponding diagrams, so that readers may compare the diagrams in this appendix directly to the system diagram. There is also a set of generic failure diagrams shown in Figures D.21 through D.24, which are used in many other failure trees but are not shown on the system diagram; these are indicated by "gen" in the black circle next to the title. We did not develop failure trees for two of the emergency response functions that are included in the system diagram: Hazmat Contain/Mitigate and Law Enforcement. We consider both functions to be outside the primary scope of this study, as described in the main body of the text.

## Failure Tree Diagram Elements

There are nine major graphical elements in each of our failure diagrams: green boxes, blue triangles with arrows leading in or out, yellow boxes, yellow circles, yellow diamonds, blue logic gates, black connecting lines, orange circles with numbers, and the previously mentioned black circles with white text. These elements follow conventions for failure diagrams established in the U.S. Nuclear Regulatory Commission's *Fault Tree Handbook* (1981). The meaning of each of these elements is explained in the next few paragraphs and summarized as a legend in Figure D.1.

Green boxes at the end of each diagram correspond to a block in the system diagram and are therefore called system failures. The text in these green boxes describes

[1] http://www.rand.org/pubs/monographs/MG994/

**Figure D.1**
**Failure Tree Legend**

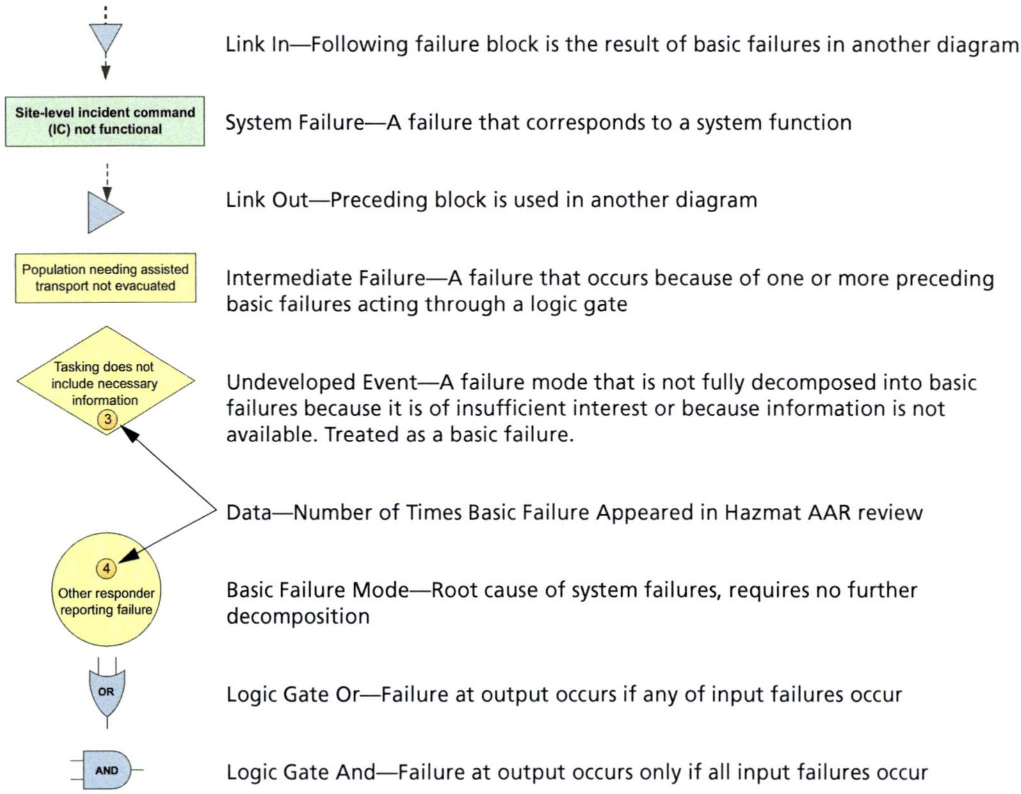


RAND *MG994-D.1*

how a failure in that block manifests functionally. For example, in diagram A (Figure D.2), "Information Received," the green box with the text "Information about scene/incident not collected, not received, or is of poor quality" describes the immediate functional consequence of a failure in the "Information Received" system block. In diagram O (Figure D.16), "Assess Resource Requirements," the green box with the text "Site incident action plan (IAP) not developed, incomplete, or incorrect for incident goals given the available resource pool" describes functionally what happens when the system function "Assess Resource Requirements (Demand Assessment)" goes poorly.

Triangles with arrows leading in or out signify a link from one diagram to another. Green boxes with arrows pointing into them show how failures in one part of the system affect other parts of the system. In each diagram, any green box with an arrow leading in matches a green box with an arrow leading out at the end of another failure tree. For example, failures in assessing the site-level resource requirements—represented by "Site incident action plan (IAP) not developed, incomplete, or incorrect for incident goals given the available resource pool" box in diagram O (Figure D.16)—

affect all site-level activities, and therefore this box shows up in the general "Resource Shortages" failure tree (Figure D.24) and in the general public communications failure and protective action trees in diagrams H and I (Figures D.9 and D.10).

Yellow boxes are intermediate failures, which group basic failures into logical classes. In diagram A, "Information Received," information input failures are divided into three classes: poor or no information provided by the public or the business at whose facility the incident occurred; poor or no information from responders on the scene; and poor or no information from automatic sensors placed prior to the incident.

Yellow diamonds and yellow circles are undeveloped and basic failures, respectively. These failures are the root cause of system failures and are the types of events and actions we looked for in the AARs. Undeveloped failures are failure types that could conceivably be broken down into more basic failures but were left aggregated because they were not as important as other basic failures in our analysis or we felt information at a more basic level would be difficult to find. For example, "Communications are incorrectly targeted," in diagram I (Figure D.10), "Protective Action Communication," could be caused by decisionmaking failures, information failures, or poor command implementation, but we felt that including that level of detail would unnecessarily complicate an already detailed failure tree.

The distinction between basic failures and undeveloped failures is somewhat arbitrary, since the emergency response system is based almost entirely on human decisions and actions rather than technological breakdowns. Use of diamonds instead of circles primarily reflects how the failure trees were expanded and condensed as we developed the trees and the coding system. Undeveloped failures can be treated as basic failures in this analysis with no loss of information.

Whether an intermediate failure occurs is determined by how it is logically connected to more basic failures. Black connecting lines show which yellow elements are linked to which blue logic gates. In diagram A, "Public/biz agent does not report well" has six contributing basic failures, which are linked with an OR gate. This means that if any one of those basic failures occurs, then there is a failure in public/business reporting of the incident. The three intermediate failures, however, are connected to the system failure "Information about scene/incident not collected, not received, or is poor quality" by an AND gate. This means that all of the intermediate failures must occur for an overall failure in information received to occur. In actual events, multiple sources often report the same incident. For example, if a fire alarm fails in a building, an employee in the building may still call 911 to report the fire. If neither occurs, then information may not reach the emergency response system in a timely manner.

The diagrams also show, in orange circles inside each of the yellow basic failure circles or diamonds, the number of times that each basic failure was found in the set of hazmat AARs. While the numbers inside the smaller orange circles support the discussion of hazmat incidents in Chapter Six, the failure trees themselves are in many cases applicable to any type of incident and emergency response.

The remaining pages of this appendix show each failure diagram and provide a brief description of its components.

## Diagram A—Information Received

Diagram A (Figure D.2) depicts failures that lead to poor reporting of incident details to the system-level incident command, or EOC. Reports can come from the public or a business at or near the site of the incident, responders on the scene, and sensing technology in place at the scene. If none of these sources provide good information, then the system will not have good information, and an information received failure occurs. While an initial report from some source is necessary to start the emergency response, information continues to come in through this tree throughout the incident.

Reporting from businesses as opposed to the general public is intended to capture reports from the operating employees and owners in an industrial accident, reports from the owners or drivers in a transportation accident, or other situations where knowing the contents and layout of a specific facility will improve the quality of the emergency response. Autonomous sensing technology at the scene may be a fire detector or security camera installed at the site location prior to the event. Information from air monitoring equipment set up at the scene by responders would most likely be reported to the on-scene commander, who would pass the information up to the system through the responder reporting branch on this diagram (shown at the bottom of the figure). Information obtained by the system command from the media is included in "Other."

## Diagram B—Establish and Operate Emergency Operations Center

Diagram B (Figure D.3) represents failures that could occur while standing up and running the system-level incident command function. (As explained earlier in the document, we use the term *emergency operations center* (EOC) to refer to any kind of system-level command function made up of facilities, technologies, people, and their roles and responsibilities rather than to just a designated EOC facility.) As with Information Received, many of the basic failures are most relevant at the beginning of the incident, but they may also be factors as the incident continues.

A well-running EOC is a function of the preestablished EOC plans and procedures, how well individuals execute those plans and procedures, the actual availability of staff, and any unanticipated disruptions that occur during the incident. Execution of plans and procedures includes the quality of facilities and technologies available for use by the EOC. Disruption of the EOC may include being overwhelmed by unnecessary people, distracted by unmanaged media, or damaged by the incident.

**Figure D.2**
**Information Received Failure Tree (A)**

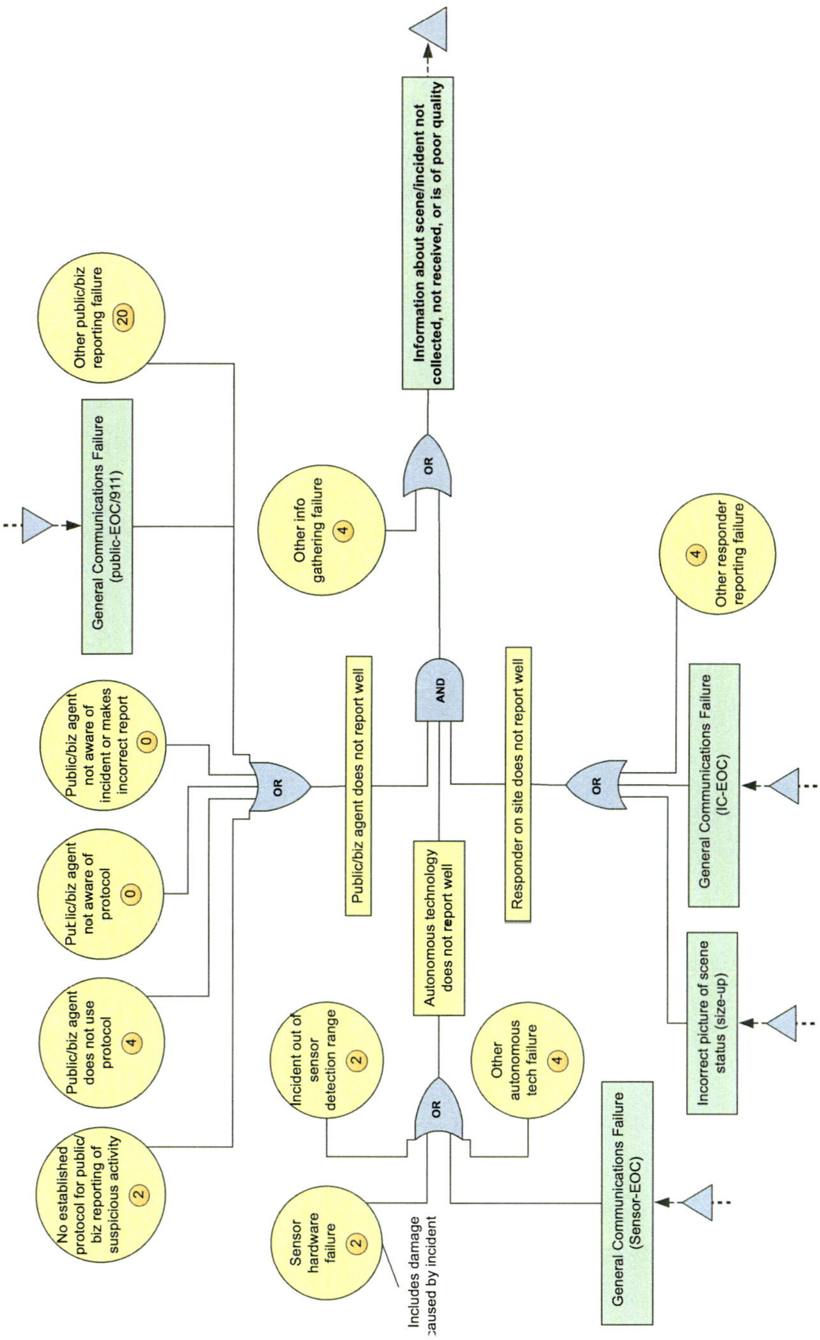


RAND *MG994-D.2*

**Figure D.3**
**Establish and Operate Emergency Operations Center Failure Tree (B)**

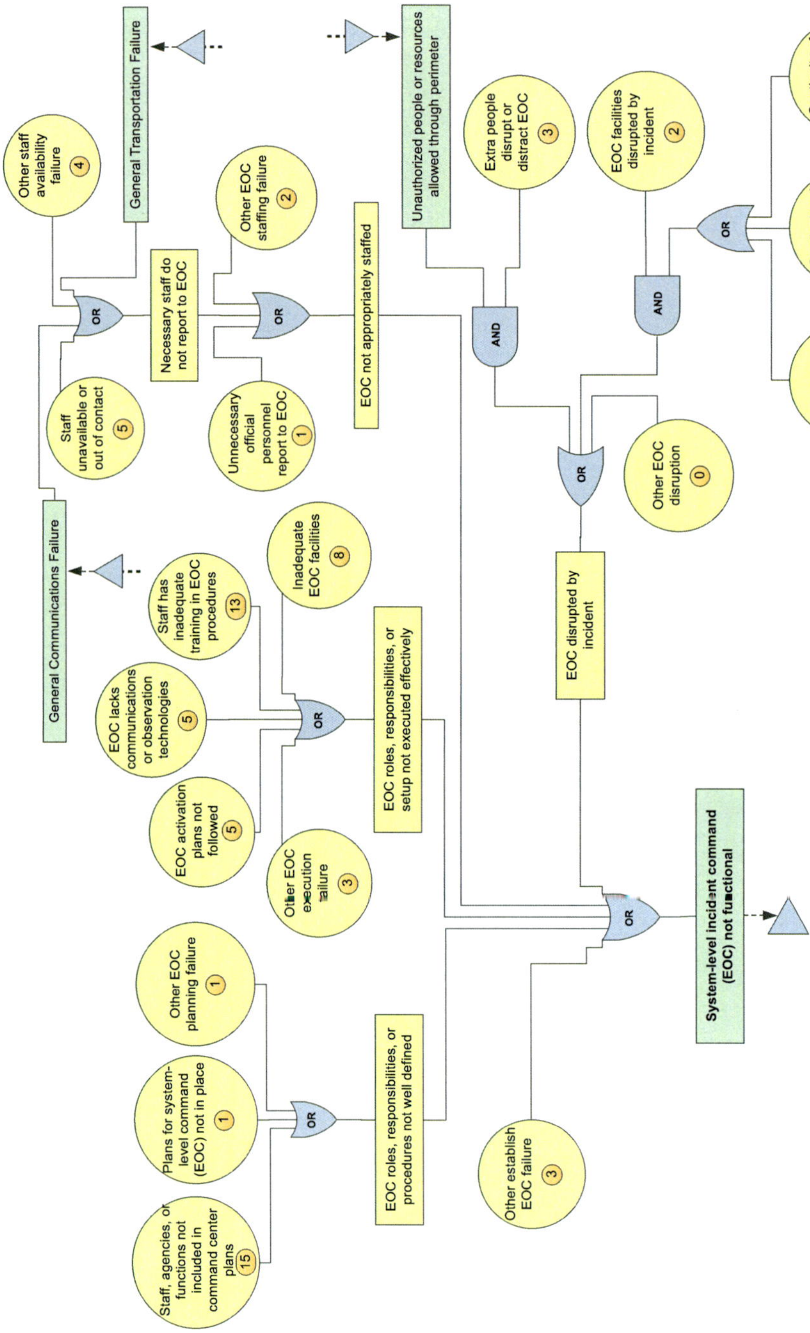


RAND *MG994-D.3*

## Diagram C—Manage System Resources

Diagram C (Figure D.4) describes how well the system-level command, or EOC, understands the resources at its disposal. It is divided into two primary branches: "Resources assumed to exist aren't available to the EOC" and "EOC does not realize resource/inventory/capability is available." A failure in either branch can lead the EOC to incorrectly understand the available resources and therefore develop plans that cannot be executed or are suboptimal in other ways.

The EOC may incorrectly assume that resources exist when they leave their expected location (physical or communication), when resources do not link to the EOC and therefore cannot be commanded, or when they simply do not arrive from outside the system (mutual aid) and the EOC assumes they are incoming. Resources may not link, or make themselves available to the EOC, because they do not know the EOC is active, they do not recognize the EOC's authority, they do not understand the incident management system the EOC is using,[2] or for other reasons. In addition, if the EOC is sufficiently dysfunctional, then resources that wish to integrate themselves in the system-level command structure may not be able to do so.

When resources do not link to the EOC, the EOC may not know that they are available. This would be the case if mutual aid or volunteer resources self-dispatched to the system but did not inform the EOC that they had arrived. This type of event is represented by the AND gate joining "Resource does not link with EOC" and "Resource self-dispatches to system." The EOC may also have a poor understanding of available resources if its personnel or inventory tracking system is incorrect, poorly executed, or nonexistent.

## Diagram D—Develop Picture of Incident Status

Diagram D (Figure D.5) represents analyses of the incident done at the system level. Both the current incident status, such as what happened and where, and the future potential incident status, such as the likely location and strength of the chemical plume in one hour, are necessary for a good response.

The quality of information available to the incident command (from diagram A) is a primary input into developing a picture of the incident status. Incoming information must also be processed such that it is available in a useful form to those doing the assessment and forecasting. Even if good information is available to the system, errors in the assessment and forecasting process may still occur. It is also possible that the system-level command will simply fail to conduct an assessment or forecast. Finally, a

---

2 We use the NIMS as shorthand for any incident management protocol that the system-level command may be using and the responding resources could be expected to know.

**Figure D.4**
**Manage System Resources Failure Tree (C)**

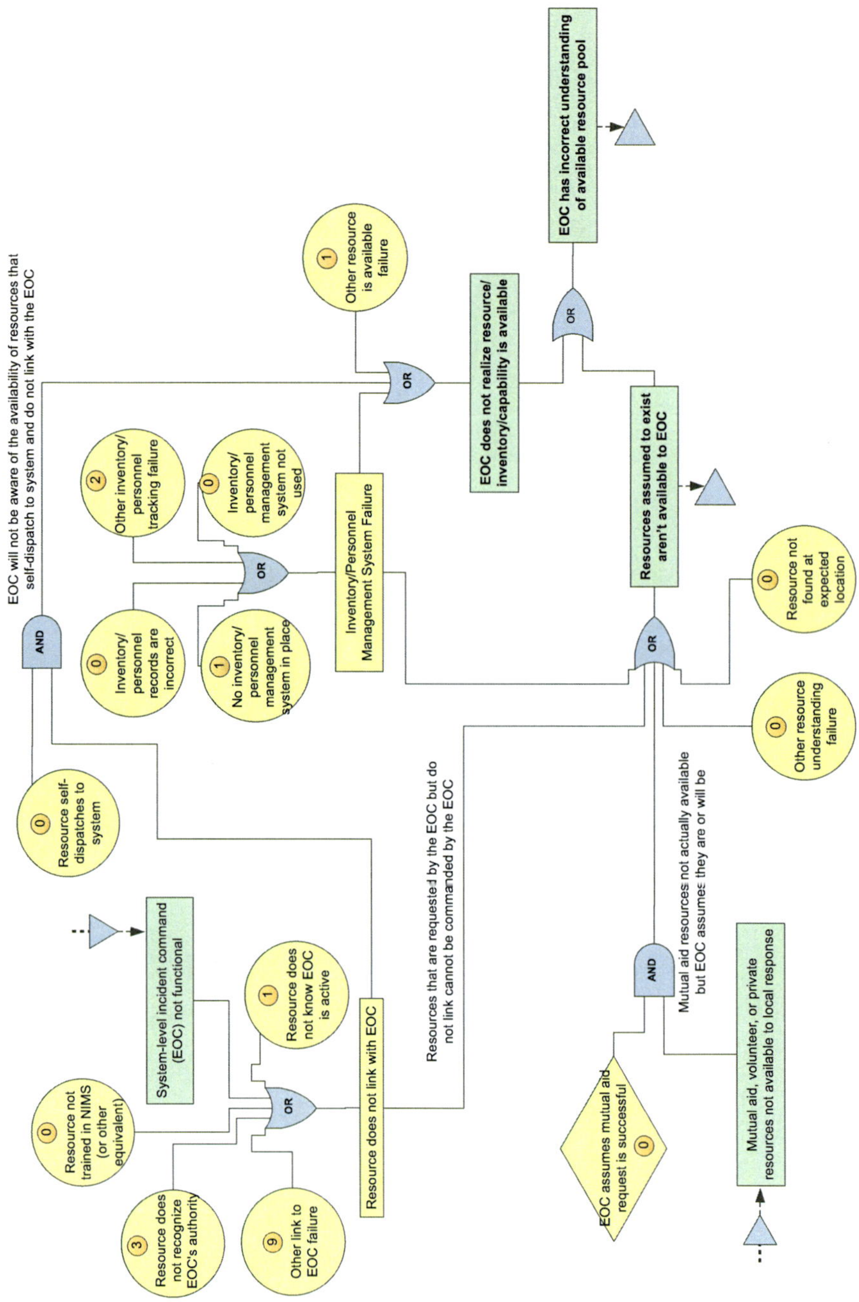


RAND *MG994-D.4*

**Figure D.5**
**Develop Picture of Incident Status (D)**

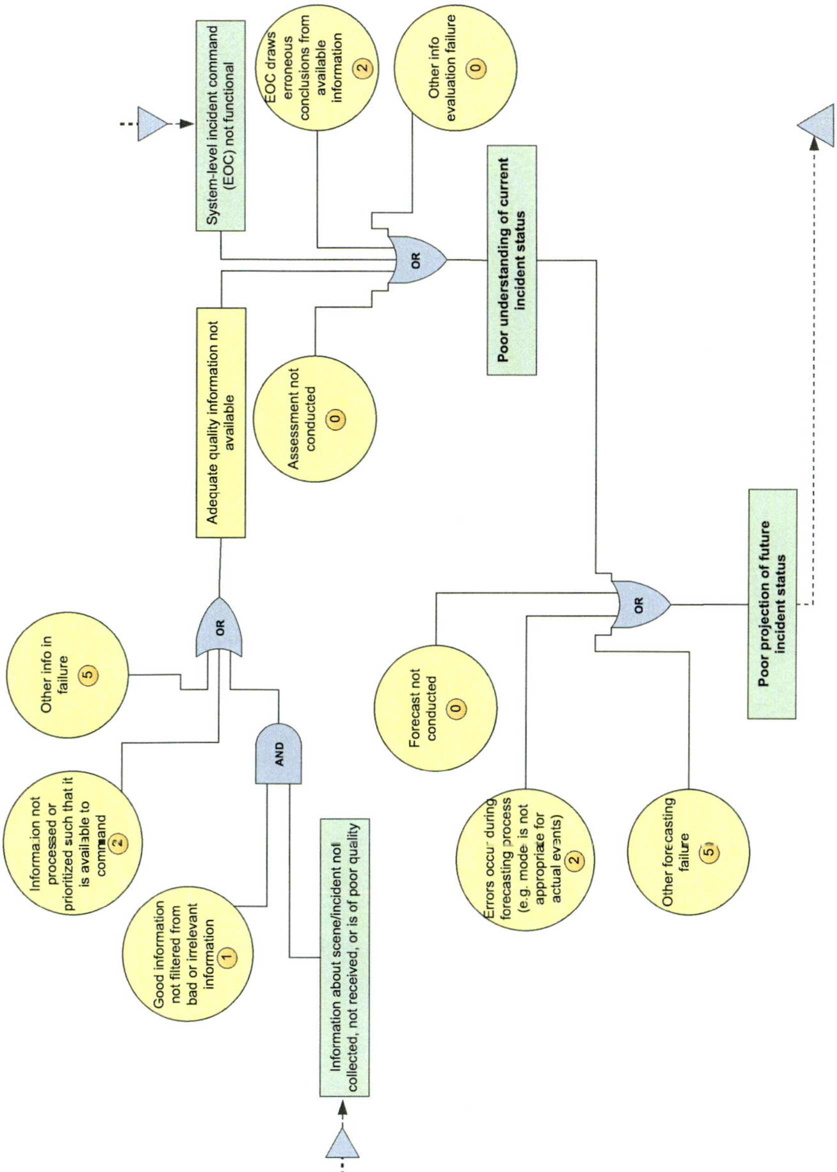


RAND *MG994-D.5*

good assessment of the incident's current status is necessary to determine an accurate future status of the incident.

## Diagram E—Dispatch Specified Resources to Site(s)

Diagram E (Figure D.6) describes potential failures to dispatch system resources to incident sites based on a system-level incident action plan (IAP). A failure in dispatching occurs if the resource listed in the IAP is not dispatched to or does not arrive at the location specified for that resource in the IAP. An IAP that is inadequate to address the actual incident is covered in diagram F.

If the EOC is not functioning well (diagram B), then dispatching may not occur. Even if the EOC is functioning well overall, other failures may cause EOC personnel to neglect their dispatch functions or contact the incorrect resource. The resource may also not be available as expected (diagram C), or general communications problems may disconnect the EOC and its target resource (see Figure D.21, the general failure tree for communications).

Even if the correct resource receives a dispatch instruction from the EOC, the dispatcher or the resource may misunderstand the instructions, the resource may choose to ignore the instructions, or the resource may not make it to its target site because of failures in the transportation system (see Figure D.22, the general failure trees for transportation and staging).

## Diagram F—Develop Desired Allocation of Resources to Sites

Diagram F (Figure D.7) depicts the process of developing a system-wide response plan to the incident. We use the term *system-level incident action plan* (system-level IAP) to represent any generic plan that assigns system resources to response sites. A high-quality system-level IAP provides a reasonable, though not necessarily perfect in hindsight, match between incident status, response goals, and available resources. The quality of the system-level IAP is a function of the quality of EOC operations (diagram B), the understanding of the actual and future incident status (diagram D), the EOC's understanding of available system resources (diagram C), and the procedures used by the EOC to develop the IAP. Even if all of these aspects function well, decisionmakers in the EOC may simply make an error (see Figure D.23, the general failure tree for decisionmaking) that leads to a poorly specified IAP.

**Figure D.6**
**Dispatch Specified Resources to Site(s) Failure Tree (E)**

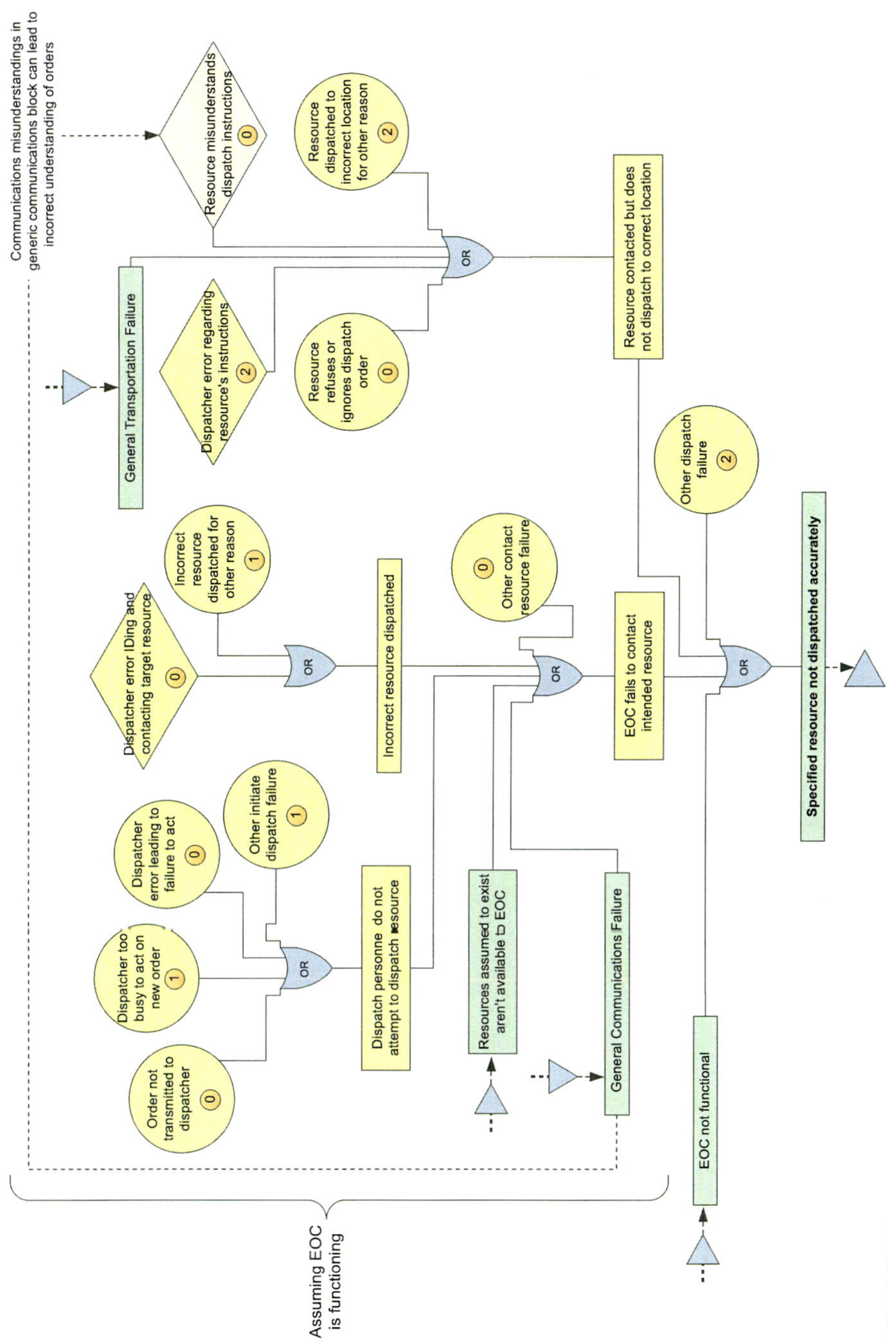


RAND *MG994-D.6*

**Figure D.7**
**Develop Desired Allocation of Resources to Site(s) Failure Tree (F)**

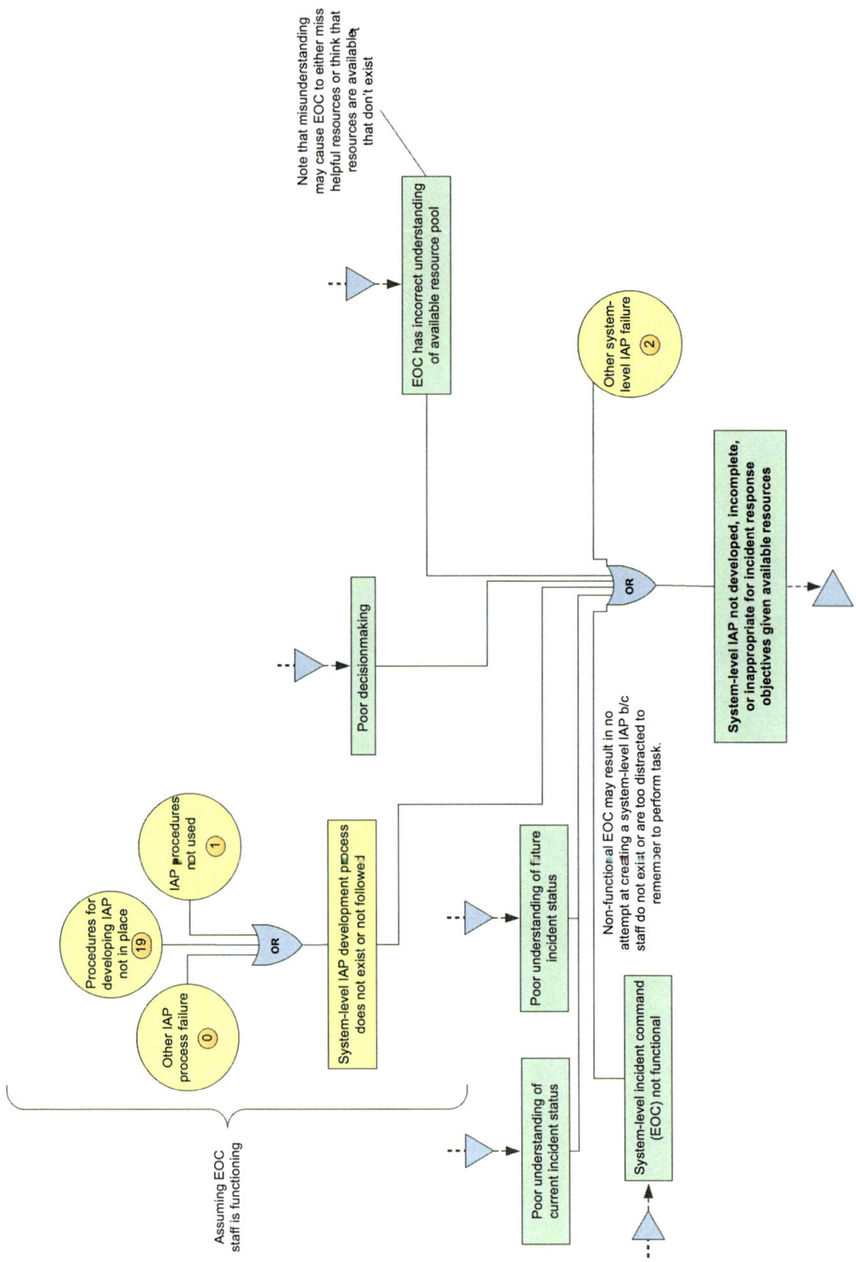


RAND *MG994-D.7*

## Diagram G—Request More Resources from Others

Diagram G (Figure D.8) depicts failures while requesting mutual aid resources. This diagram is similar to the diagram for dispatching resources from the system to sites (diagram E). A failure to request and receive mutual aid resources may occur because the EOC did not request aid, because the resource was requested but not sent by the other organization, or because the resource could not travel correctly from its home location to the incident site. The EOC may not request aid because it is not functioning well (diagram B), because within a well-functioning EOC responsible staff neglect to request the resource or ask for the wrong resource, or because the EOC does not know how to contact a particular resource. General communications failures may also prevent a request for aid from reaching the other organization (see Figure D.21, the general failure tree for communications). If the resource is requested, the request may be denied because a mutual aid agreement is not in place, because even with an agreement no one is available in the other organization to authorize the resource, or because the other organization has no resources to spare. If a mutual aid resource is dispatched, it may be prevented from reaching the incident scene due to general failures in the transportation system (see Figure D.22, the general diagram for transportation) or because of incorrect instructions regarding where to go.

## Diagram H—General Population Communications

Diagram H (Figure D.9) depicts potential failures that occur when attempting to distribute official messages about the incident to the general population. This function is distinct from instructions to evacuate or shelter (diagram I). Failures in general population communications can lead to a confused public or members of the public placing themselves in harm's way.

Site and system public information officers (PIOs) are usually responsible for implementing general population communications. If either the system-level or the site-level IAP (diagram F or diagram O, respectively) do not provide the PIO with adequate instructions and support, then general population communications may not occur or may be incorrect.[3]

We divide general population communications failures into six intermediate failure categories. Most of these intermediate failures have only two contributing basic failures: failures in the IAPs and all other causes. These six types of failures are grouped into two branches. First, the population may receive the message but not behave in the

[3] If the PIO role is not designated at the site or system level, this failure will show up in diagram L, "Establish and Operate Site-Level Incident Command," or in diagram B "Establish and Operate EOC." Poorly functioning command in turn affects the site- and system-level planning.

**Figure D.8**
**Request More Resources from Others Failure Tree (G)**

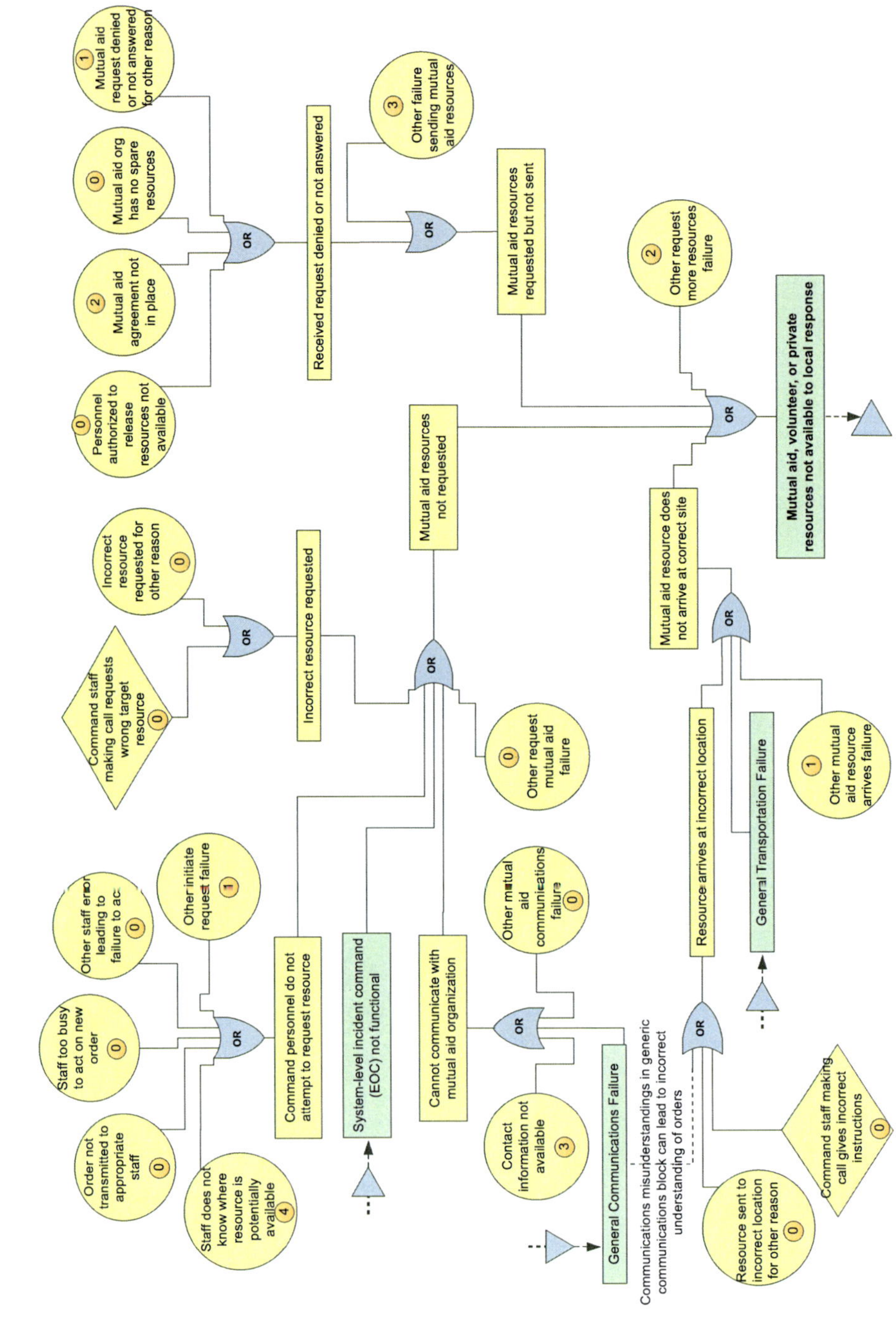


RAND *MG994-D.8*

**Figure D.9**
**General Population Communications Failure Tree (H)**

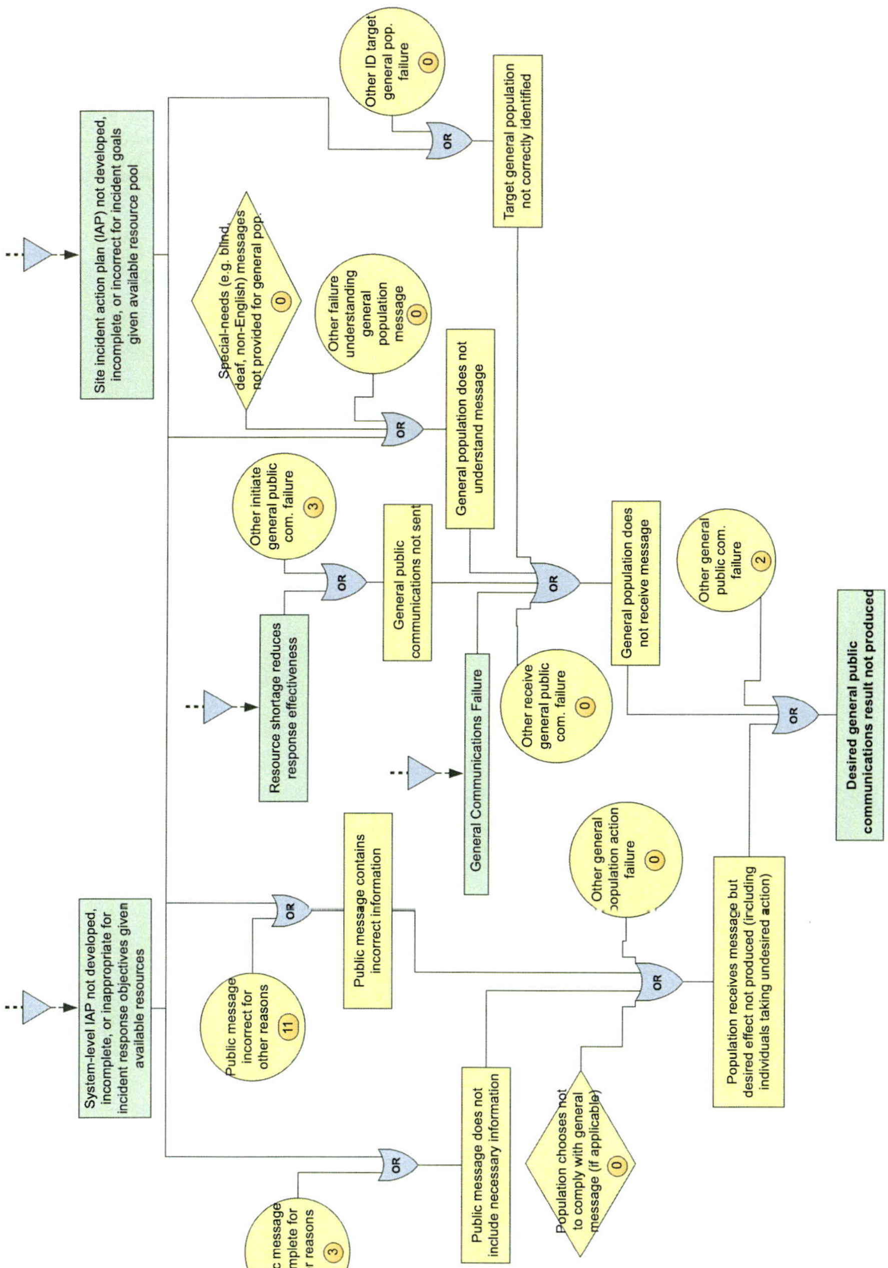


RAND *MG994-D.9*

manner command would like, either because the message has incorrect information,[4] the message does not contain sufficient information, or because the population chooses not to comply.[5] Second, the population may not receive the message, which could occur if command does not target the correct general population, if the public does not understand the message, or if command simply neglects to send the message.

## Diagram I—Protective Action Communications

Diagram I (Figure D.10) is very similar to diagram H, "General Population Communications," but instead deals with specific instructions to evacuate or shelter in place. Protective action communications failures may result in two kinds of system failures: (1) part of the affected population may not take protective action and (2) an unaffected population may take action that adversely impacts the response, such as evacuating needlessly, thereby increasing traffic along evacuation routes.

Diagrams H and I have nearly identical elements, but the elements have slightly different meanings. Instead of linking to failures in the PIO function as in diagram H, in diagram I, site- and system-level incident action planning failures refer to failures in organizing evacuation or shelter-in-place plans. Also, targeting the correct population to receive the protective action is split into two parts: identifying the affected population and correctly using the communications system to reach them.

Just as with general population communications (diagram H), protective action communications failures are divided into six intermediate failure categories and grouped into two branches. First, the population may receive the message but not take action because either the message has incorrect information,[6] the message does not contain sufficient information, or the population chooses not to comply. Second, the population may not receive the message, which could occur if command does not target the correct population, if the public does not understand the message, or if command simply neglects to send the message.

---

[4] This diagram does not account for corrections to the message. If an incorrect, incomplete, or unintelligible message was distributed at any time in the response, even if it was later corrected, the failure is recorded according to the codes established in this diagram.

[5] For example, in a wildfire, people in areas not affected by the fire may drive to the fire site in order to see what is going on, despite instructions to the contrary.

[6] As with general population communications, the protective action communications diagram does not account for corrections to the message.

**Figure D.10**
**Protective Action Communications Failure Tree (I)**

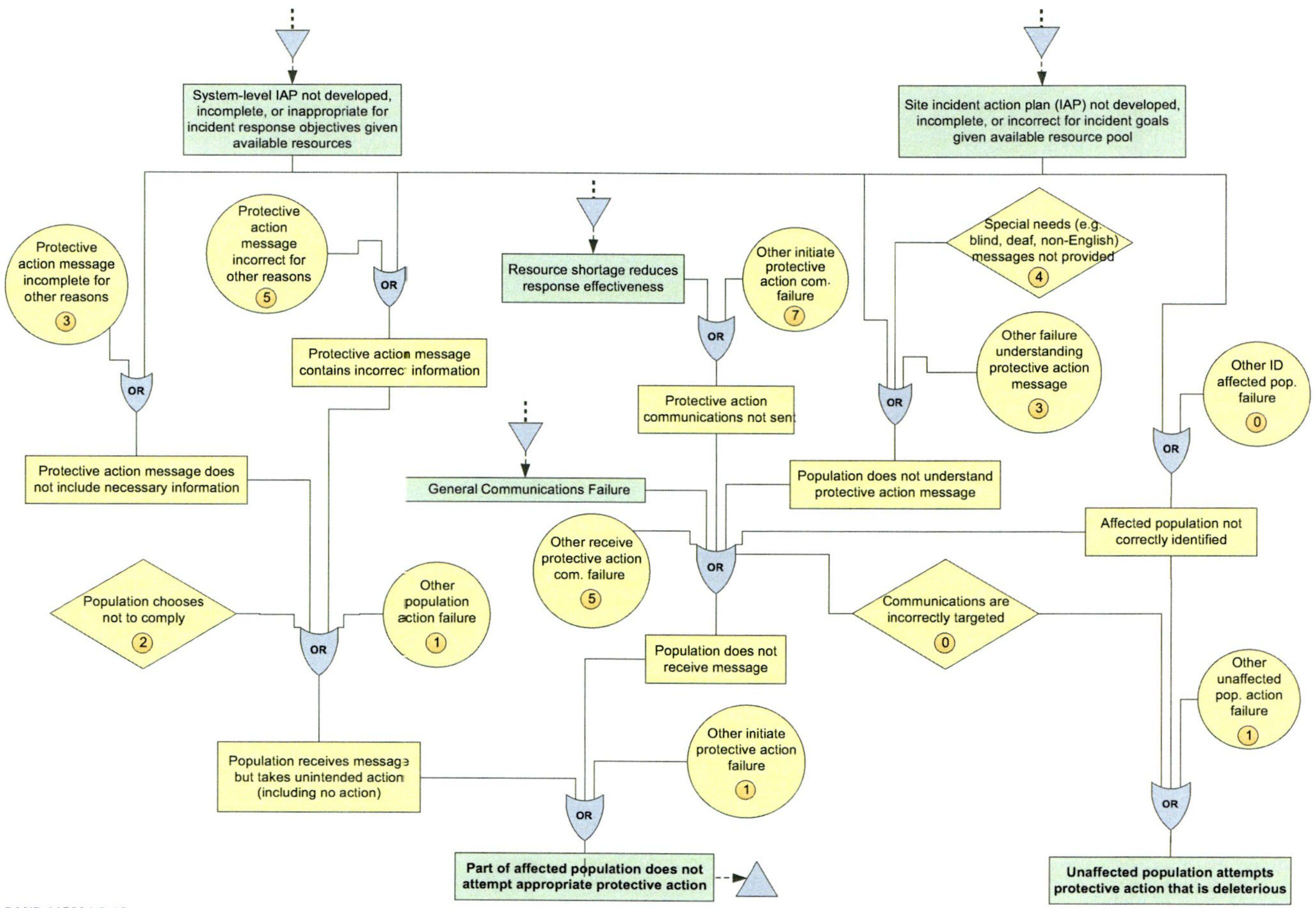


RAND *MG994-D.10*

## Diagram J—Evacuation and Shelter-in-Place

Diagram J (Figure D.11) depicts failures in evacuations or shelter-in-place actions. An evacuation or shelter-in-place action is successful to the extent that it protects the affected population from the incident hazard. If the population does not receive or chooses not to act on a protective action message from command (diagram I), then no evacuation or sheltering will occur. Assuming there is a real threat to their life and health, populations that do not act will be harmed. If the population does receive the protective action message and begins to comply, several other failures may still result in injury to the population.

Shelter-in-place failures are the most straightforward. If a shelter-in-place is called, people may be injured if sheltering is not an effective defense against the threat. All other potential failures in sheltering are included in the other category.

If an evacuation is called, the population may not be able to evacuate or they may be injured while complying with the evacuation order. General transportation failures or the incident itself may prevent some people for evacuating.[7] Special needs populations may need assistance, and if enough shelters are not opened then the evacuating population may have no safe place to stage while waiting for the evacuation to be lifted. Finally, people may be harmed by criminal activity or panic if the there is insufficient security.

## Diagram K—Responder Safety and Health

Diagram K (Figure D.12) shows the failure tree for the response function that tracks and manages the safety and health of other responders. We portray two types of failures in responder safety and health: responder performance may be degraded or responders may be injured. For optimum responder performance, the responder safety and health function must be tasked through the site-level IAP (diagram L) and staffed sufficiently (see Figure D.24, the general failure tree for resource shortages).

The injury failure branch is primarily concerned with injuries that occur in a hazmat event. The basic failures therefore cover different reasons why a responder may not be using appropriate PPE and yet be tasked to work in a hazardous area. Responders may also be injured because their performance is degraded. We mainly considered performance impacts due to fatigue, which may occur because the system does not have sufficient resources to relieve responders or because responders ignore orders to

[7] Stephen Brittle's article critiquing the Graniteville train crash response (Brittle, no date) notes that workers at a factory near the chlorine spill had difficulty starting their cars because the chorine and the humid air had created an acid that destroyed the ignition. (Note that we did not include Brittle's article in the AAR review because we already had many documents written by the response agencies involved.)

**Figure D.11**
**Evacuation and Shelter-in-Place Failure Tree (J)**

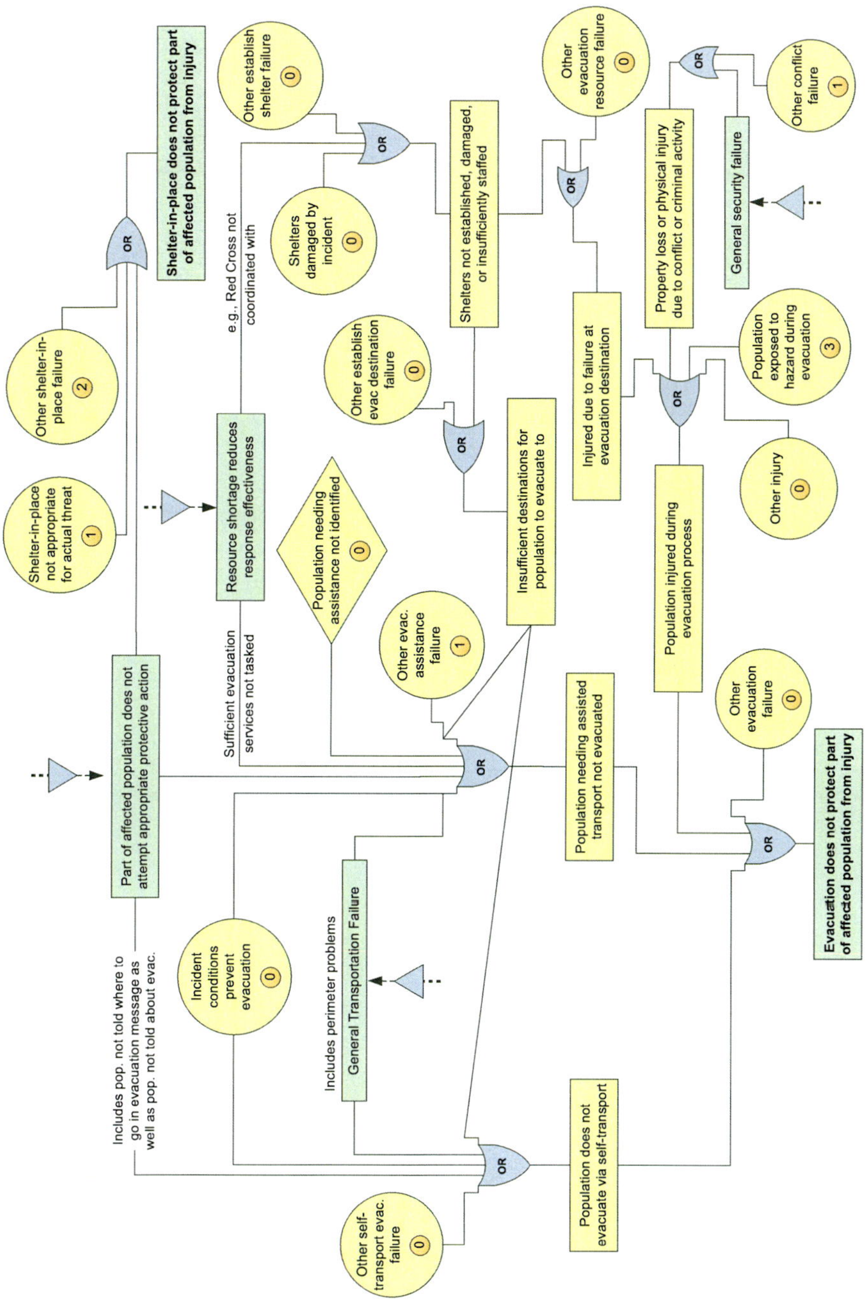


RAND *MG994-D.11*

**Figure D.12**
**Responder Safety and Health Failure Tree (K)**

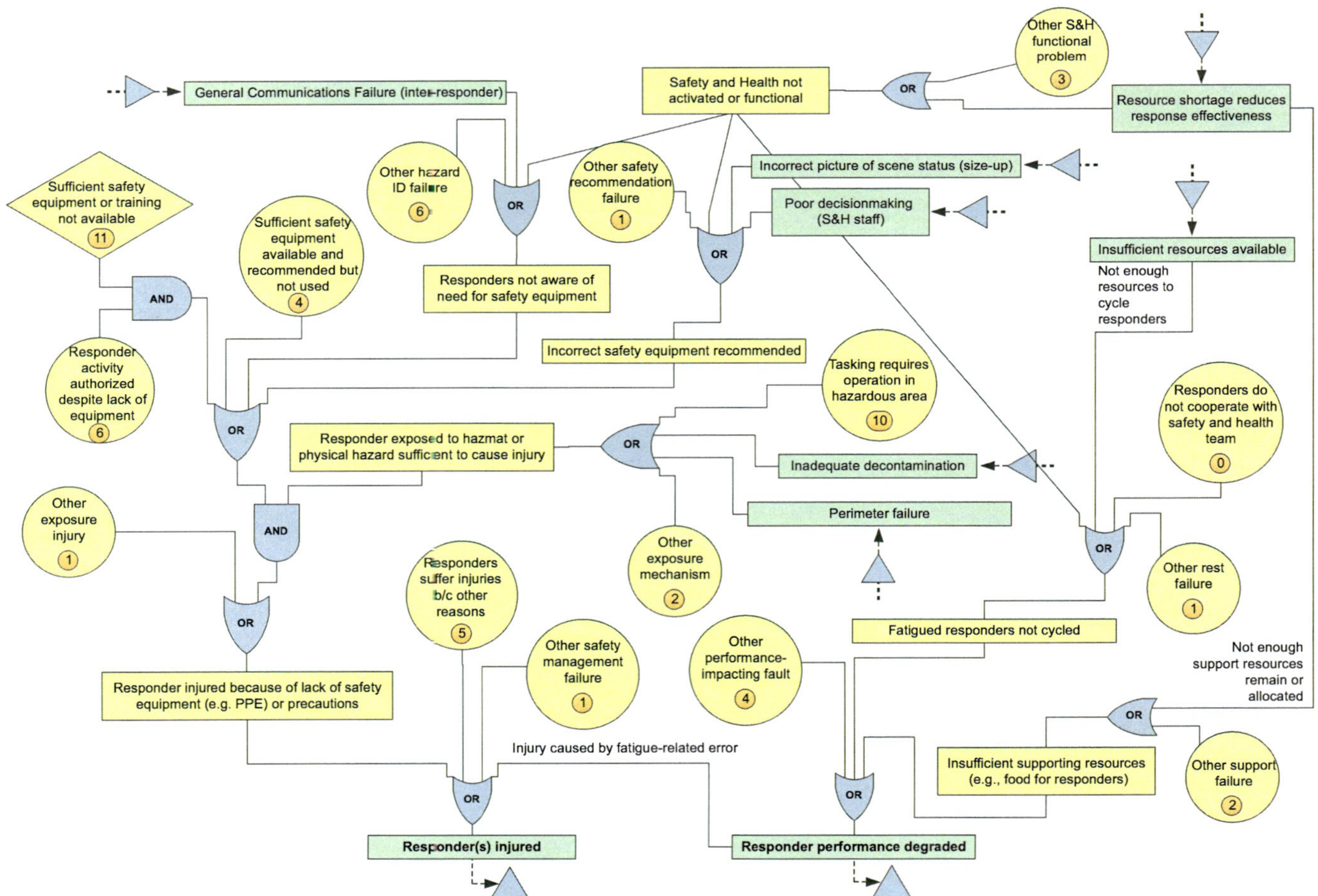


RAND *MG994-D.12*

rest. Insufficient food, water, or other supporting resources can also reduce responder performance.

## Diagram L—Establish and Operate Site-Level Incident Command

Diagram L (Figure D.13) describes the events that can lead to a poorly functioning site-level command. If the site-level incident command is not functioning well, many of the site-level response functions may also fail. We use the term *incident command* (IC) to represent any generic site-level command structure. The IC failure diagram is nearly identical in structure to the system-level EOC diagram (diagram B). The IC may function poorly because the IC plans and procedures were not well established prior to the incident, IC plans were not implemented effectively, the IC was not appropriately staffed, or the IC was disrupted by the incident.

## Diagram M—Size-Up Scene

Diagram M (Figure D.14) describes failures that may prevent evaluation of the incident on-site. Some information about the incident, such as details from a public call to 911, may flow down from the system level to the site level. Failures in system-level information collection and analysis may result in poor information at the site level as well. Responders at the scene may also be tasked with directly collecting information about the incident. If they are not tasked, if they perform their task poorly, or if the necessary technology, such as air monitoring equipment, is not available, then the site-level command may not receive necessary information about the scene.

## Diagram N—Manage Site Resources

Diagram N (Figure D.15) is the site-level version of diagram C, "Manage System Resources." As with the system-level diagram, site-level resource management failures can cause the IC to miss resources that are actually available or to assume resources are available that are not.

The IC may incorrectly assume resources exist when they leave their expected location (physical or communications), when resources do not link to the IC and therefore cannot be commanded, or when they are not successfully dispatched and the IC assumes they are incoming. Resources may not link, or make themselves available to the IC, because they do not know the IC is active, they do not recognize the IC's authority, they do not understand the incident management system the IC is using, or

**Figure D.13**
**Establish and Operate Site-Level Incident Command Failure Tree (L)**

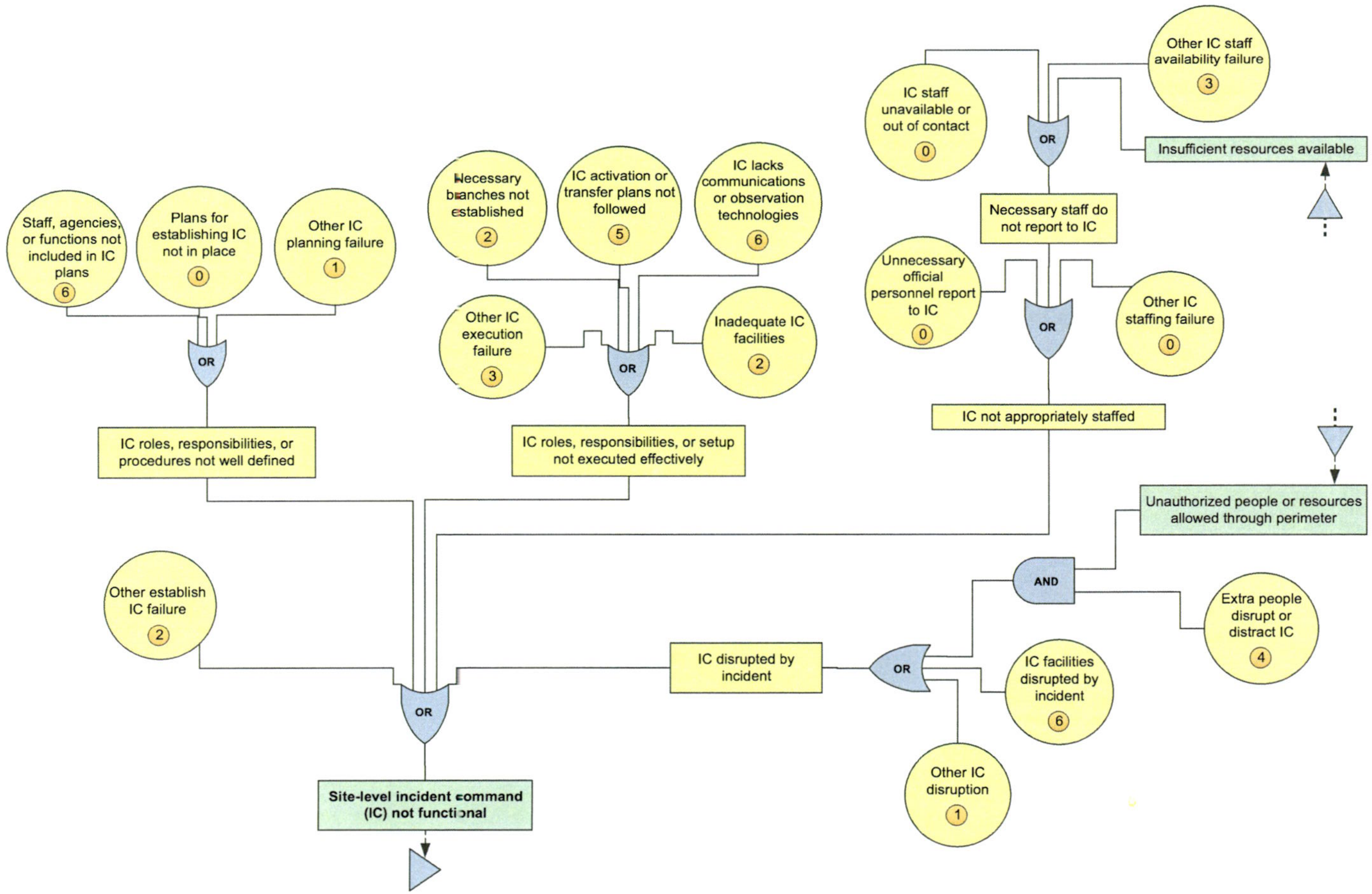


RAND *MG994-D.13*

**Figure D.14**
**Size-Up Scene Failure Tree (M)**

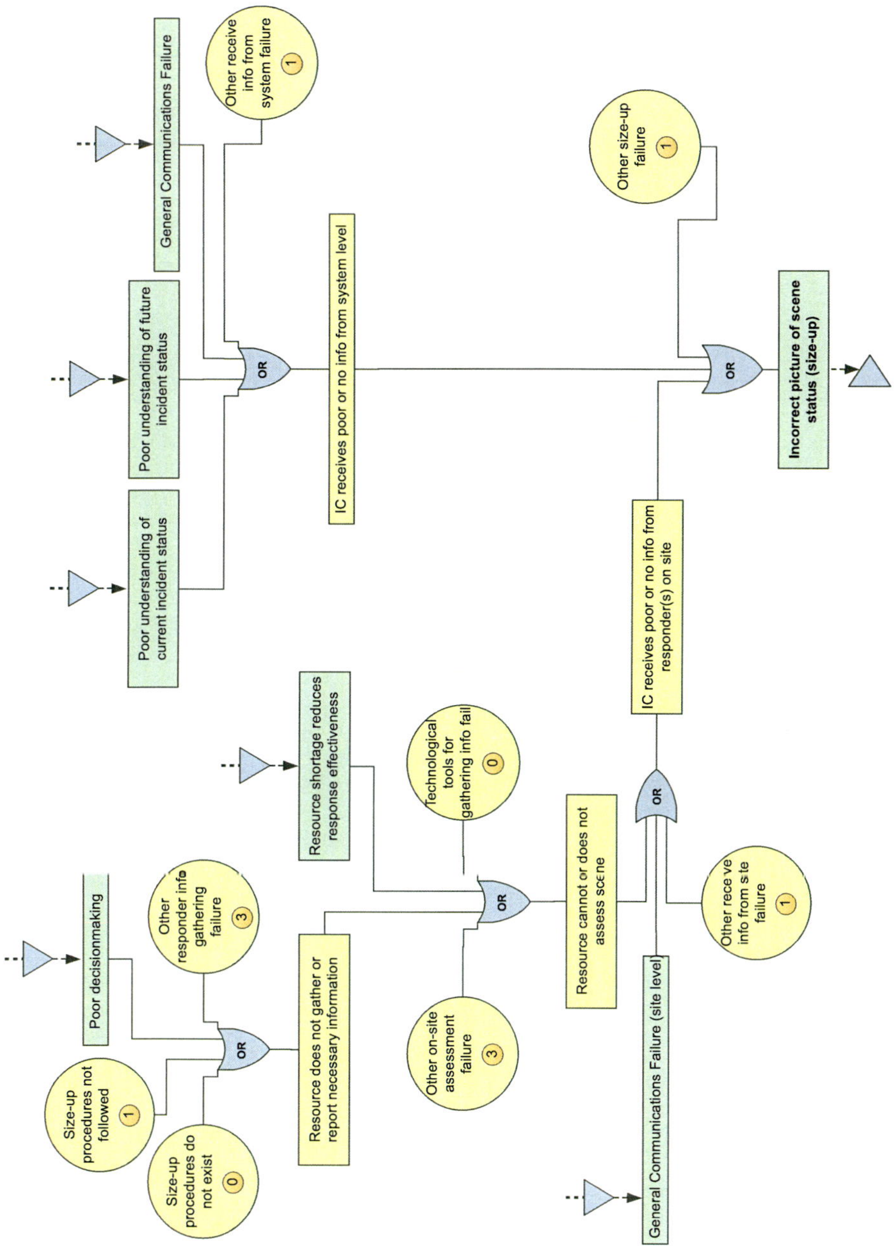


RAND *MG994-D.14*

**Figure D.15**
**Manage Site Resources Failure Tree (N)**

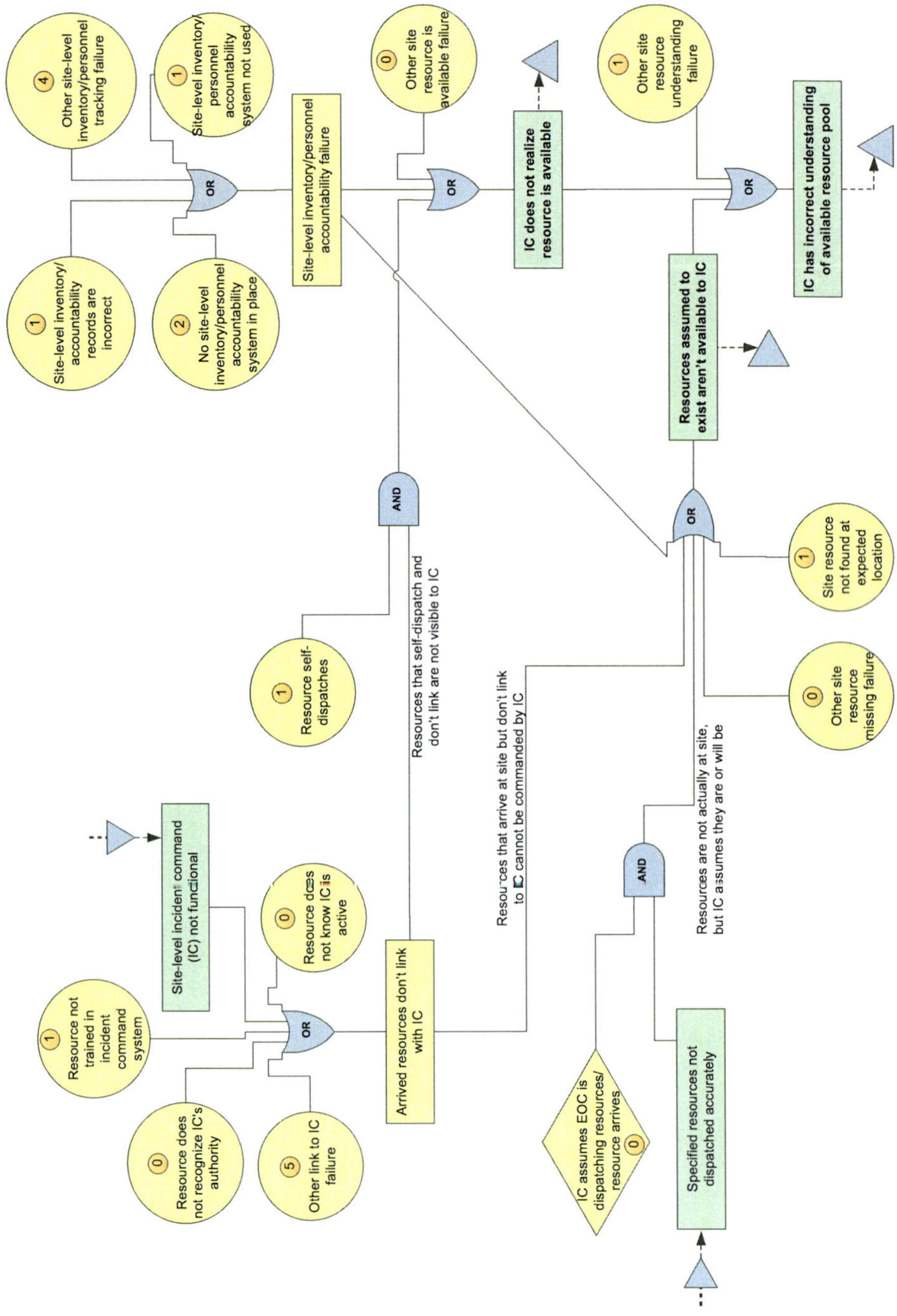


RAND *MG994-D.15*

for other reasons. In addition, if the IC is sufficiently dysfunctional, then resources that wish to integrate themselves in the IC structure may not be able to do so.

When resources do not link to the IC, the IC may not know that they are available. This would be the case with resources that self-dispatch to the system but do not inform the IC that they have arrived. This type of event is represented by the AND gate joining "Resource does not link with IC" and "Resource self-dispatches." The IC may also have a poor understanding of available resources if its personnel accountability or inventory tracking system is incorrect, poorly executed, or nonexistent.

## Diagram O—Assess Resource Requirements

Diagram O (Figure D.16) is the site-level version of diagram F, "Develop Desired Allocation of Resources to Site(s)." This diagram describes failures in developing the site incident action plan (IAP). As before, *IAP* is intended to mean any site-level response plan that matches resources to tasks. The site IAP may be incorrect, given the incident and the available resources, if the site-level command is not functioning well (diagram L), has poor information about the incident (diagram M), or has a poor understanding of available resources (diagram N). In addition, there may be a failure in the site-level IAP if appropriate procedures are not followed by the IC in developing the IAP or if the IC personnel make a decisionmaking error (see Figure D.23, the general failure tree for decisionmaking).

## Diagram P—Task Resources According to IAP

Diagram P (Figure D.17) shows possible failure modes that lead to a mismatch between the site IAP and the resources actually sent to perform response functions. It is similar to diagram E, "Dispatch Specified Resources to Site(s)," at the system level.

If the IC is not functioning well (diagram L), then tasking may not occur. If the IC is functioning well overall, the resource may not be available as expected (diagram N), or general communications problems may disconnect the IC and their target resource (see Figure D.21, the general failure tree for communications).

If the correct resource receives a tasking order from the IC, either the command staff or the resource may misunderstand the instructions, the resource may choose to ignore the instructions, or the resource may not be able to implement the instructions due to incident conditions, transportation failures (see Figure D.22, the general failure trees for transportation and staging), or because the resource is blocked by the property owner. The resource may also be lacking the specific equipment or training needed to perform the task.[8]

---

[8] Lack of equipment or training was the most common basic failure mode in the full set of AARs and in the hazmat AAR subset. We found 56 instances of this failure overall and 25 instances in hazmat incidents. See Figure 6.1 and diagram Q,R.

**Figure D.16**
**Assess Resource Requirements Failure Tree (O)**

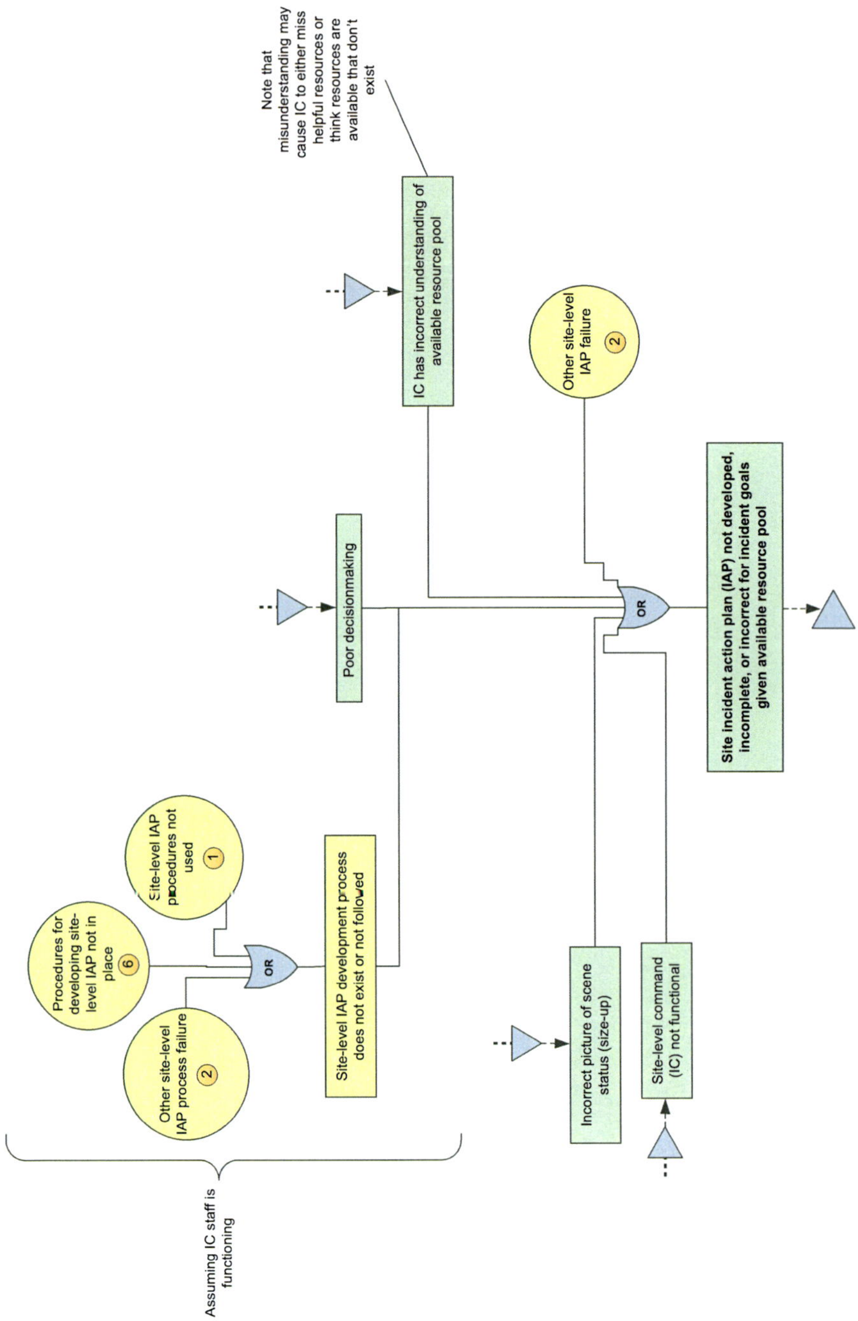


RAND *MG994-D.16*

**Figure D.17**
**Task Resources According to the Incident Action Plan Failure Tree (P)**

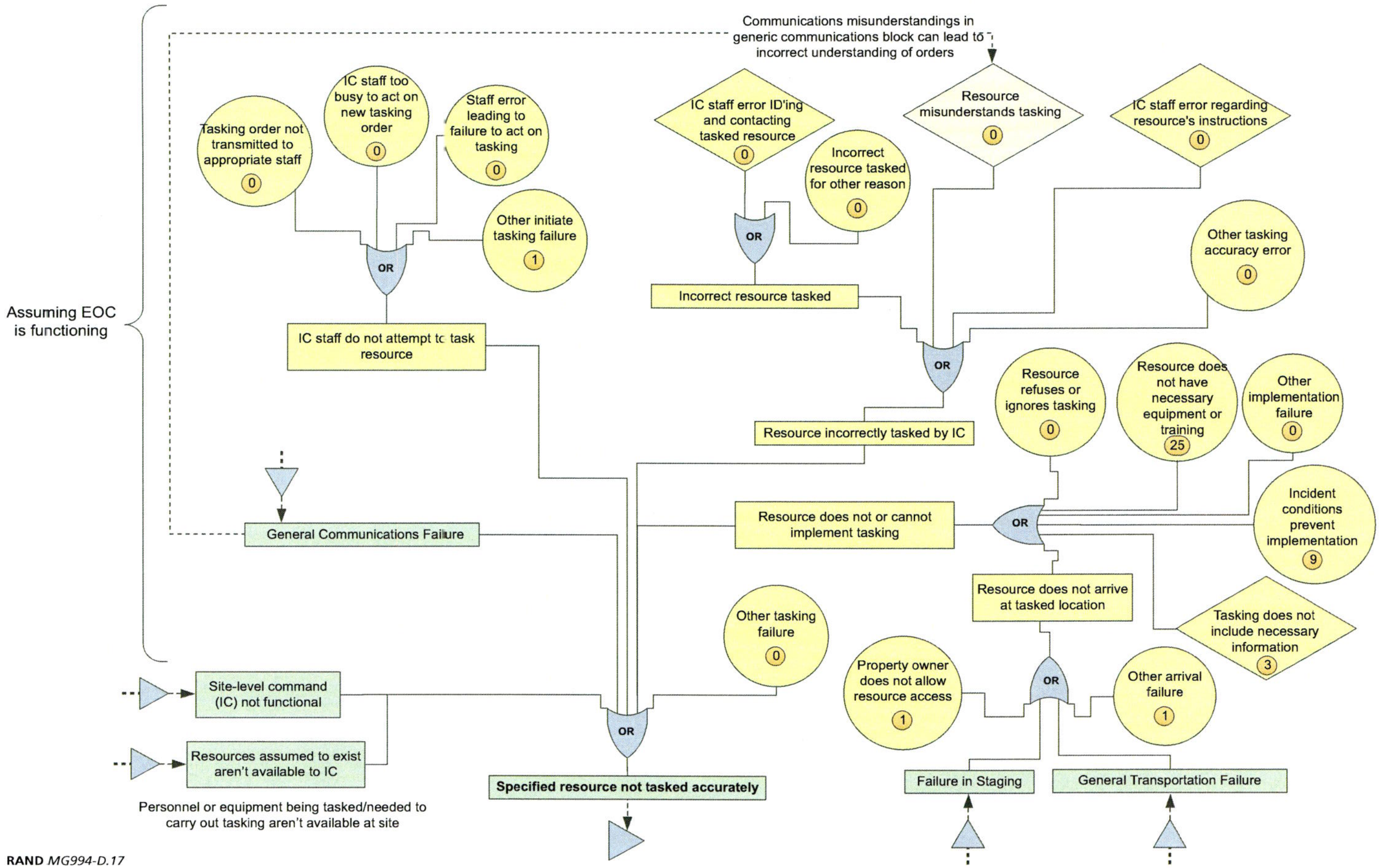


RAND *MG994-D.17*

## Diagram Q,R—Site Security and Perimeter

Diagram Q,R (Figure D.18), "Site Security and Perimeter," covers two functional branches in the response system model and shows failures that lead to poor implementation of the perimeter or any other general security failure at the incident site(s). These functions are generally performed by law enforcement. Perimeter and security failures can impact evacuations (diagram J), responder safety and health (diagram K), medical treatment and transport of victims (diagram T), general transportation and resource staging (see Figure D.22, the general failure diagram for transportation and staging), and crowding at the site and system command (diagrams B and L).

The perimeter may fail in two ways: authorized people may be blocked from the perimeter or unauthorized people may be allowed through the perimeter. We characterize blocking authorized people to be the result of a decisionmaking error (see Figure D.23, the general failure tree for decisionmaking) by the perimeter staff or an "other" reason. Allowing unauthorized people through the perimeter may be the result of a decisionmaking error or the result of incomplete perimeter implementation. The perimeter may be incomplete because of resource shortages (see Figure D.24, the general failure tree for resource shortages), including a poor IAP, or because of "other" reasons. We assume security failures are the result of perimeter failures, insufficient security staff, or "other" reasons.

## Diagram S—Victim identification and Retrieval

Diagram S (Figure D.19) shows failure modes that affect the first stage of helping victims of the incident. Failures in victim identification and retrieval mean that victims are left at the incident scene or victims are not decontaminated before coming in to contact with other responders or civilians. To successfully retrieve victims, sufficient responders must be tasked to the function (see Figure D.24, the general failure tree for resource shortages), those responders must know where the victims are (largely a function of the quality of incident information in diagrams D and M), and they must be able to access the victims' locations. In order to safely remove victims from the hot zone in a hazmat incident, responders must be able to appropriately decontaminate the victims.

## Diagram T—Medical Treatment and Transport

Diagram T (Figure D.20) shows failure modes that impact the second stage of helping victims of the incident. Medical treatment and transport failures result in not treating serious casualties and transporting them to the hospital or not treating and releasing non-serious casualties. To successfully treat and transport victims, first the victims

**Figure D.18**
**Site Security and Perimeter Failure Tree (Q,R)**

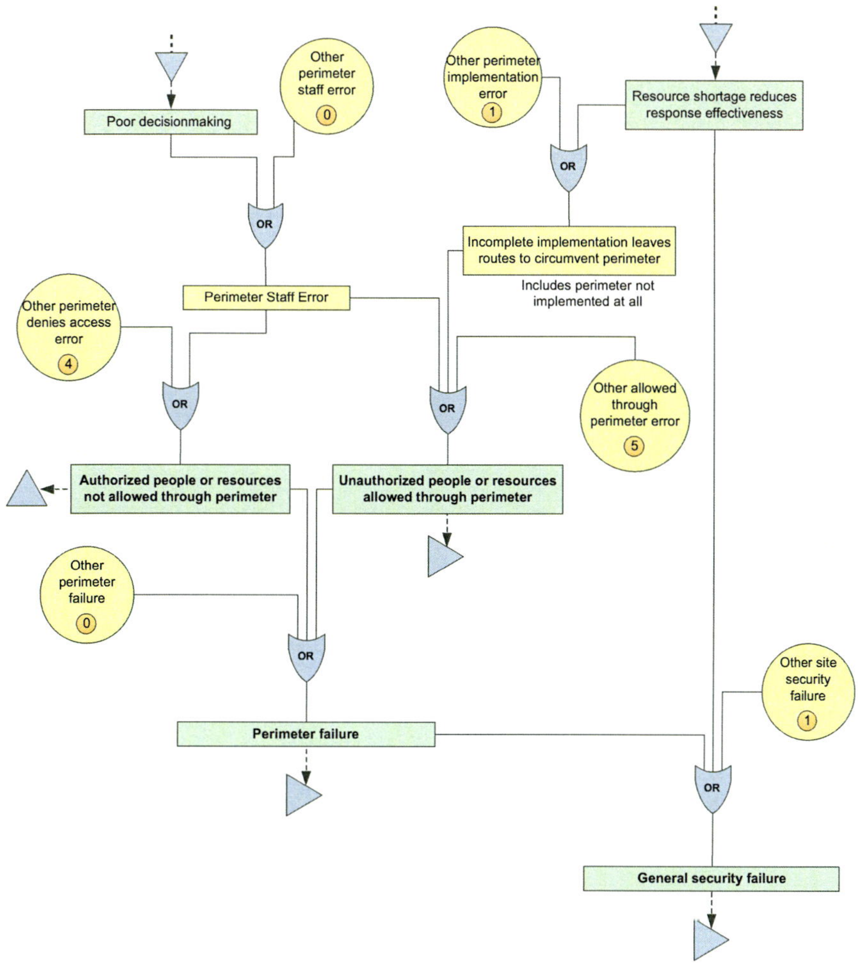


RAND *MG994-D.18*

must be retrieved from the scene (diagram S). Next the victims must be triaged (or treatment assigned) to determine the level of medical attention they require. Following treatment assignment, the victims are treated if necessary and then released on their own, or transported to the hospital. When victims with serious injuries are successfully transported from the scene, appropriate medical facilities must be available to receive the victims from the medical treatment and transport function to be complete.

**Figure D.19**
**Victim Identification and Retrieval Failure Tree (S)**

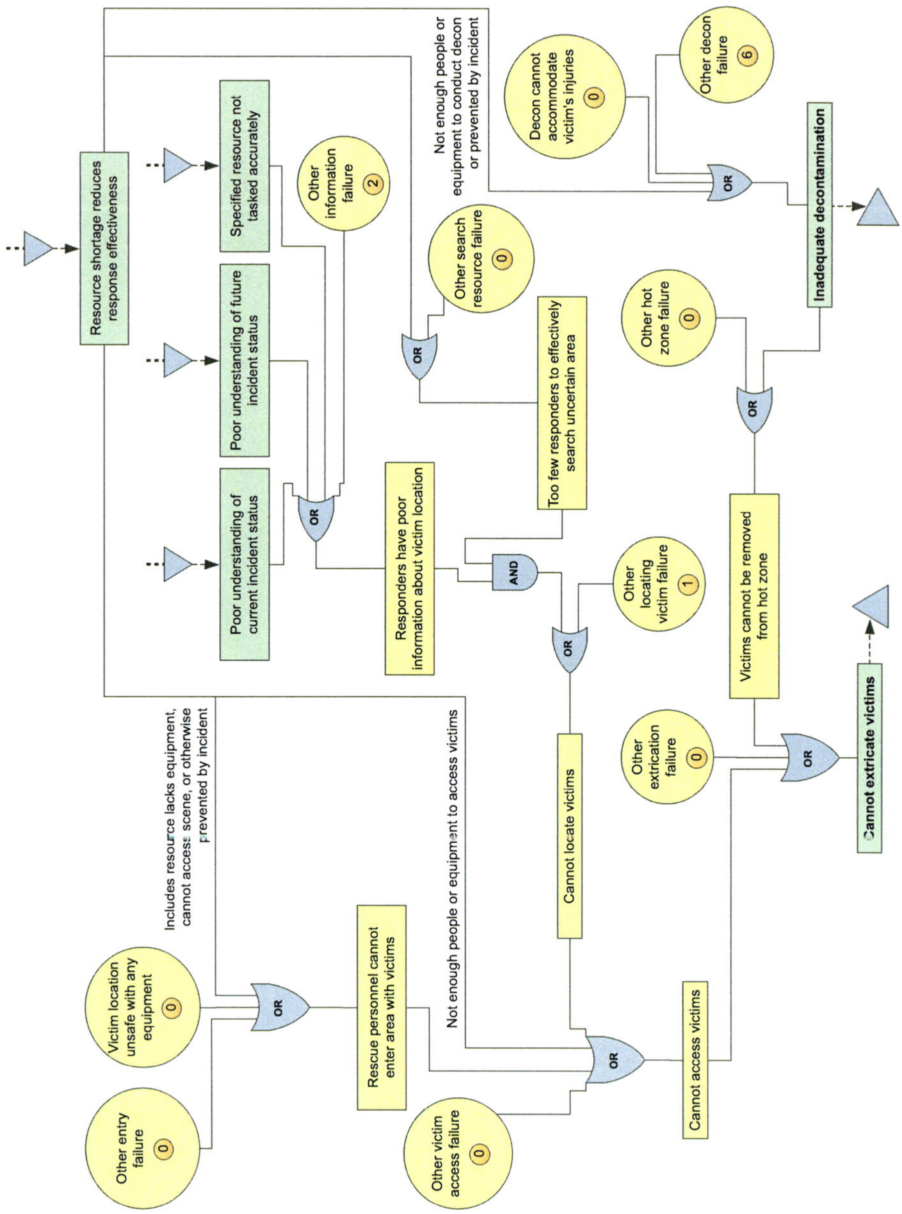


RAND *MG994-D.19*

**Figure D.20**
**Medical Treatment and Transport Failure Tree (T)**

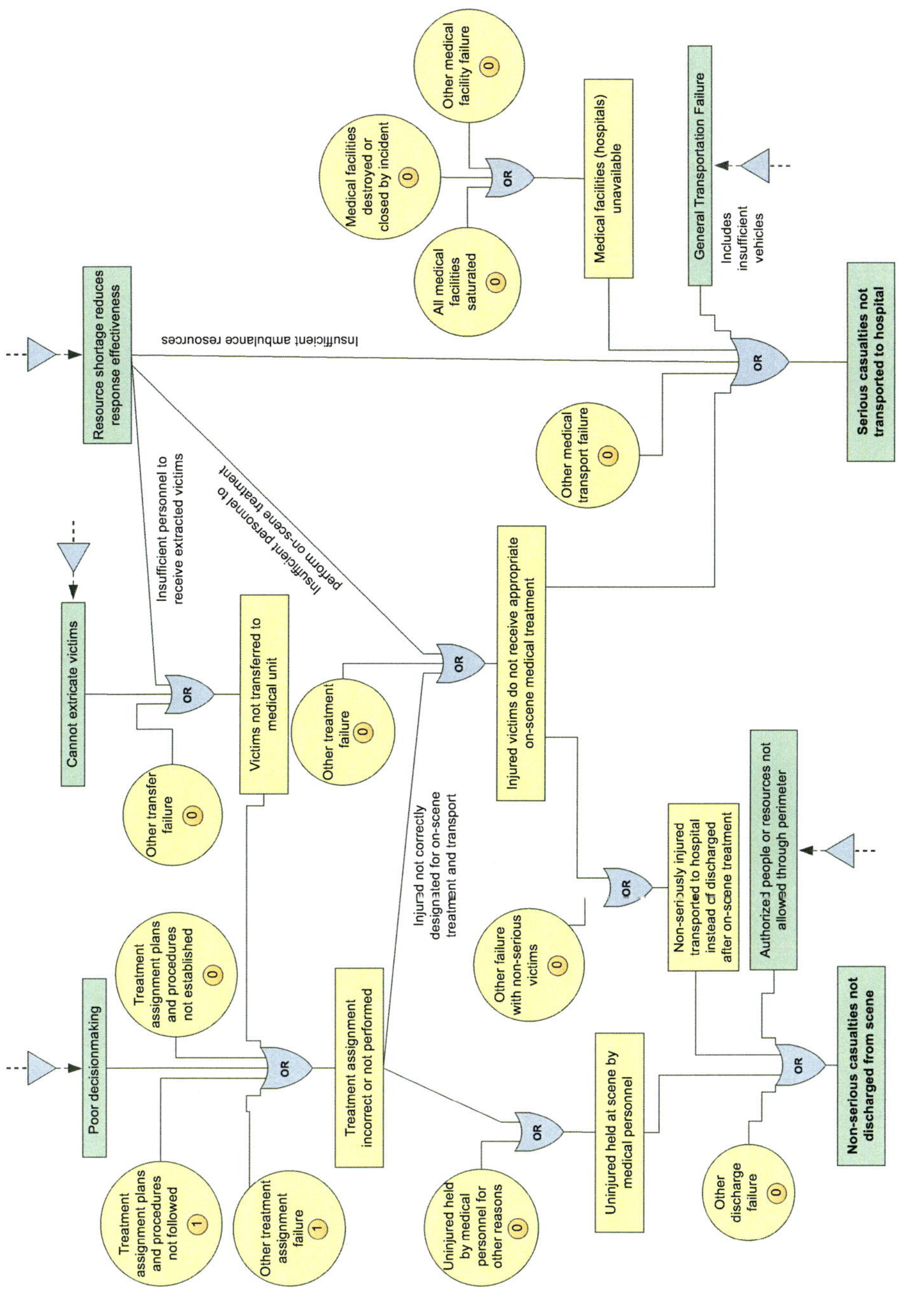


RAND *MG994-D.20*

## Diagram General—Communications

Figure D.21, the general failure tree for communications, impacts many of the other response functions. While the failure mode shown in this diagram is general and applies to any type of communications failure, in practice analysts need to keep track of what type of communications—public to responder, command to responder, responder to responder on-site—is actually affected by the failure. Otherwise, communications problems may seem more widespread than they actually are. For example, failure in public phone lines is unlikely to impact radio communications between responders.[9]

Communications failures are divided into four categories. First, the technology may fail, including backup options, or responders may not know how to use the technology. Second, communications procedures may fail, such as EOC staff not knowing which response organizations are on which radio frequencies. Third, the message may be unintelligible because of static, noise, or other interference, or the message may be poorly formulated and subject to misinterpretation. Fourth, the communications targets may simply not answer because they have walked away from their desks, are busy with other tasks, or don't notice the page (not expanded in the diagram).

## Diagram General—Transportation and Staging

Figure D.22, the general failure trees for transportation and staging, displays two related general areas that affect response operations. First, transportation failures can impact the response any time a unit needs to move from one location to another. As with communications failures, the analyst should note which response activities a transportation failure is actually impacting to avoid double counting.[10] Transportation failures may be due to traffic delaying travel, the incident damaging the transportation infrastructure, insufficient vehicles to move resources, or perimeter failures. Second, staging failures affect the efficiency of site operations. We divide staging failures into basic failures that result in an overcrowded staging area, incident damage to the staging area and resources within, and poor management of the staging area.

## Diagram General—Decisionmaking

Figure D.23, the general failure tree for decisionmaking, is mainly used to simplify the structure of other failure tree diagrams—such as developing the IAP, deciding on who

---

[9] This lesson was learned toward the end of this exploratory project, and therefore is not well implemented in our results.

[10] This lesson was learned toward the end of this exploratory project, and therefore is not well implemented in our results.

**Figure D.21**
**Communications Failure Tree**

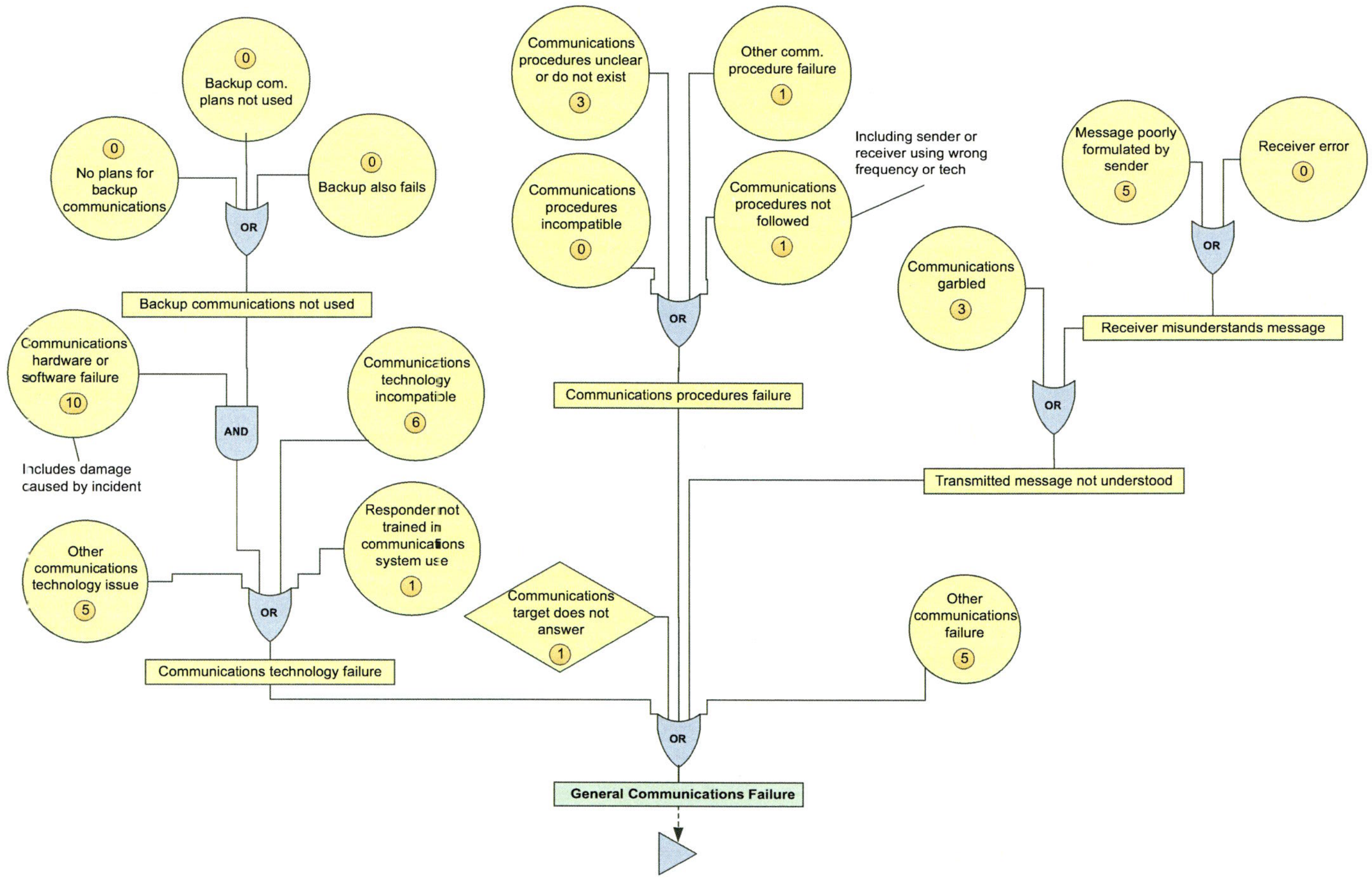


RAND *MG994-D.21*

**Figure D.22**
**Transportation and Staging Failure Trees**

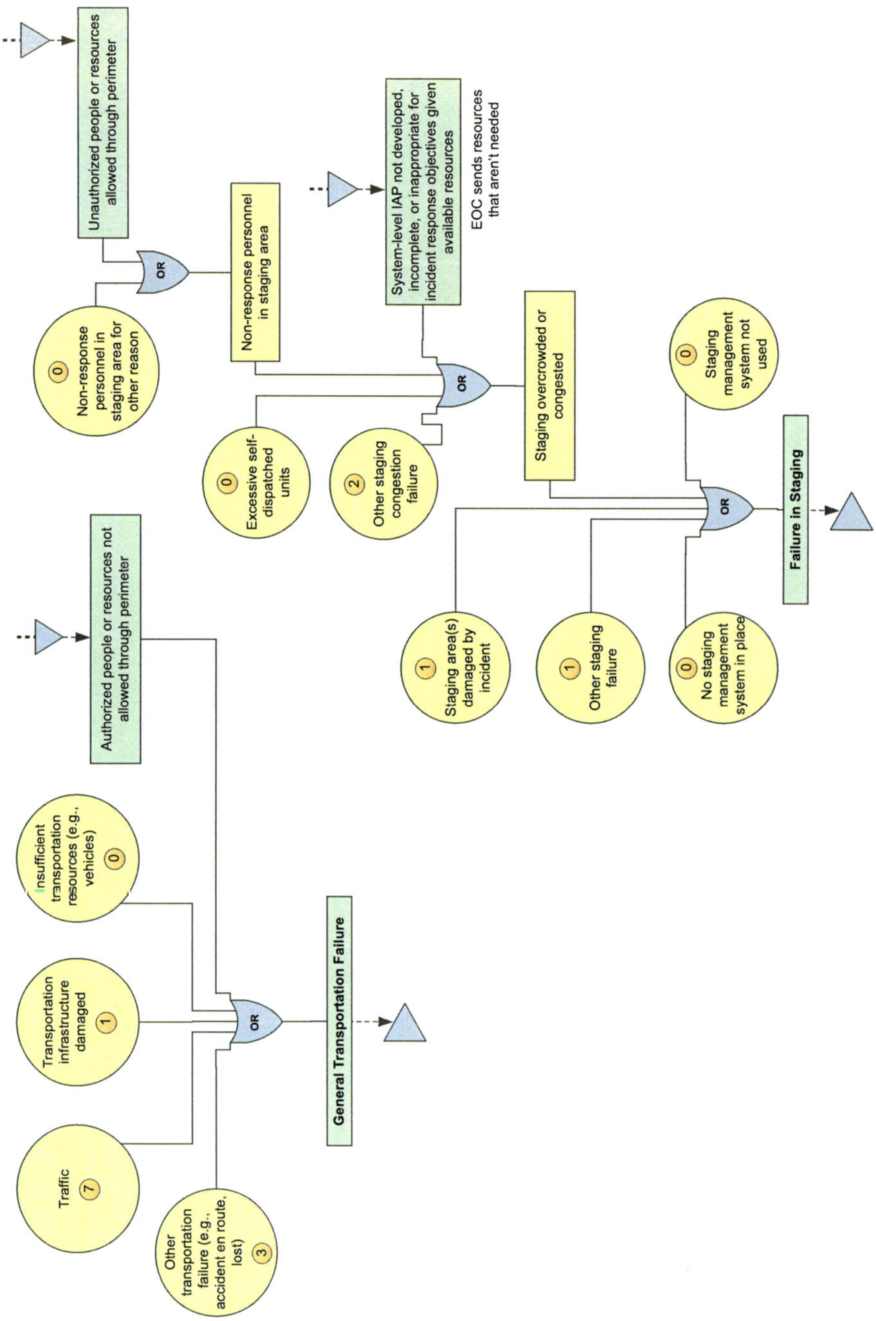


RAND *MG994-D.22*

**Figure D.23**
**Decisionmaking Failure Tree**

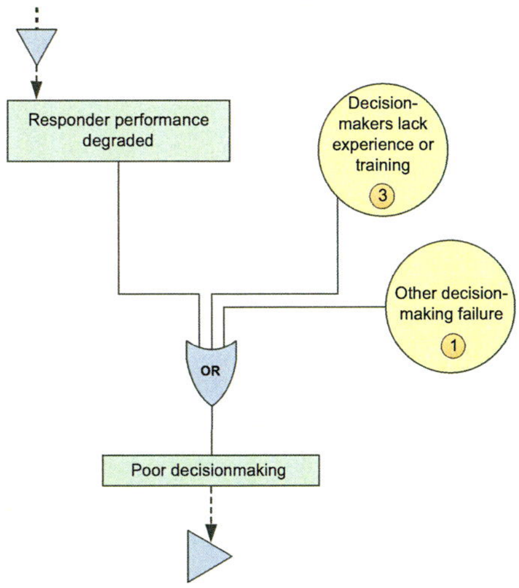


RAND *MG994-D.23*

to admit through the perimeter, and assigning treatment categories to victims—that require responders to make a decision. We depict two main causes of decisionmaking failures: poor training and degraded performance (diagram K).

## Diagram General—Resource Shortages

Figure D.24, the general failure tree for resource shortages, summarizes all of the types of failures that can lead to insufficient resources in some response tasks. If there are too few resources available to the system, it may not be possible to fully staff response functions. There may be too few resources if the system-level command fails to plan for (diagram F), request (diagram G), or dispatch (diagram E) sufficient resources, if those dispatch and requested resources do not arrive at the incident sites (diagrams G and E), if the necessary equipment is damaged, or if the pool of responders is depleted due to injury or fatigue. If the system has sufficient resources, too few responders may be assigned to tasks due to failures in the site-level IAP (diagram O) or due to failures in tasking (diagram P).

**Figure D.24**
**Resource Shortages Failure Tree**

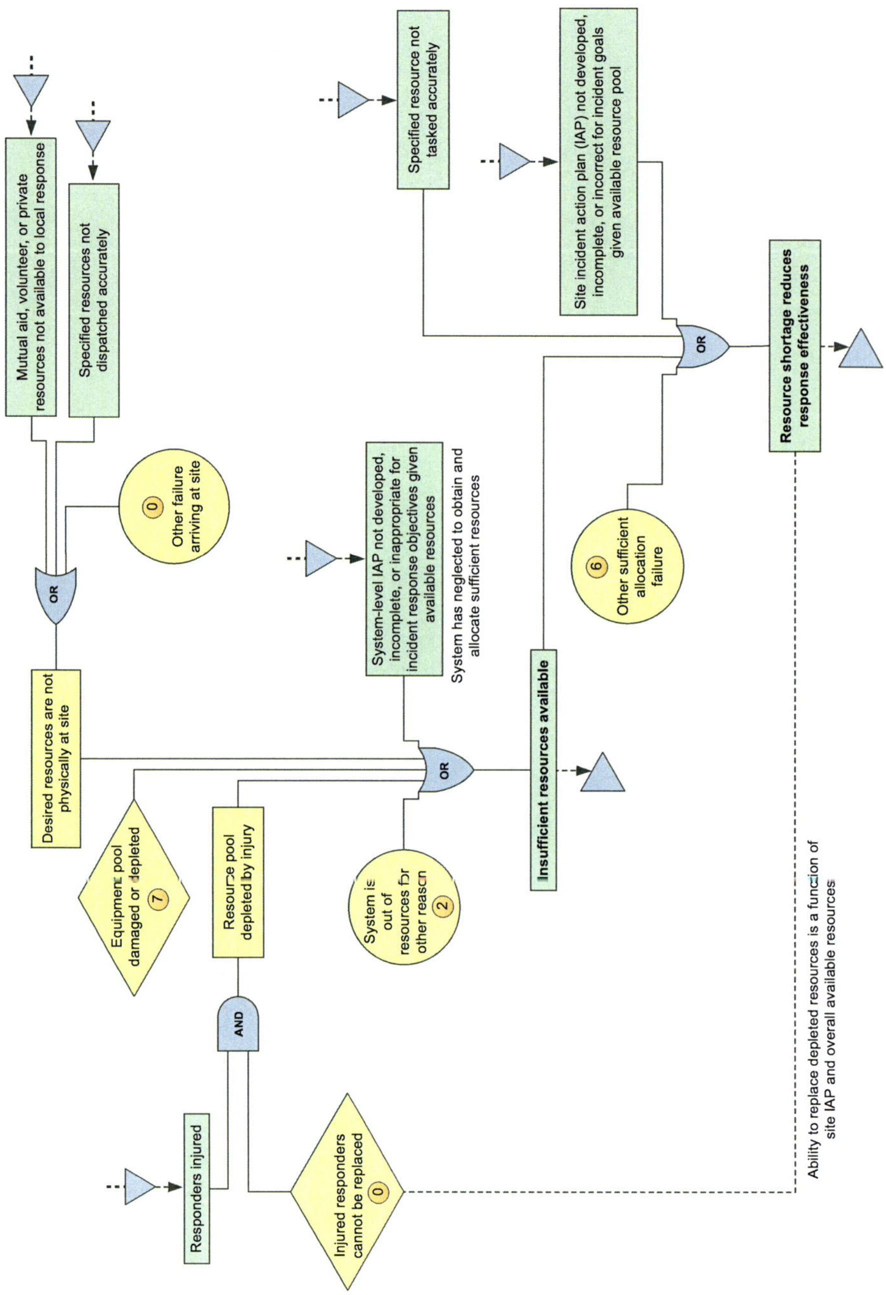


RAND *MG994-D.24*

APPENDIX E

# Counts of Failure Modes Identified per Analyzed After-Action Report

**Table E.1**
**Counts of Failure Modes Identified by Incident**

| Incident | Failure Modes Identified |
|---|---|
| Nisqually earthquake | 213 |
| 2007 San Diego County firestorms | 153 |
| Graniteville, South Carolina, train crash | 68 |
| Taft, Louisiana, chemical tank explosion | 48 |
| B'nai B'rith biological threat | 48 |
| Santiago fire | 47 |
| Topanga fire | 46 |
| 2007 Washington and Oregon windstorm | 44 |
| MFG Chemical, Inc., toxic chemical vapor cloud release | 43 |
| Volusia County terrorism rail exercise | 41 |
| Arlington, Virginia, tanker fire | 38 |
| Oklahoma City bombing | 37 |
| Burlington Northern train derailment | 36 |
| Imperial Sugar Dixie Crystal plant fire | 31 |
| Westley Tire fire | 28 |
| Chemical fire in Apex, North Carolina | 22 |
| Flint Township Industrial Plastics fire | 20 |
| 2002 Winter Olympics | 19 |
| Henderson, Nevada, liquefied chlorine gas leak | 18 |
| Baltimore Tunnel train derailment | 15 |
| DPC Enterprises, Festus, Missori, chlorine release | 14 |
| Nebraska City Tire Recycling Facility fire | 13 |
| DuPont train derailment exercise | 13 |
| City of Alamosa, Colorado, salmonella outbreak | 12 |

**Table E.1—Continued**

| Incident | Failure Modes Identified |
|---|---|
| Nanticoke Metal Processing Plant fire | 12 |
| Battle Creek Complex | 10 |
| Little General Store, Inc., propane explosion | 10 |
| Indianapolis–Marion County storms, April 2, 2006 | 10 |
| DPC Enterprises, Glendale, Arizona, chlorine release | 9 |
| Springfield, Massachusetts, swimming pool chemical plant fire | 9 |
| Tacoma, Washington, chlorine release | 8 |
| Georgia-Pacific hydrogen sulfide poisoning | 7 |
| Explosion at Isotec biochemical facility | 6 |
| Weld County tornados | 6 |
| Technic, Inc., vent collection system explosion | 6 |
| San Simeon earthquake | 6 |
| Alberton Canyon, Montana, chlorine rail car derailment | 5 |
| Indiana State Fairgrounds | 5 |
| Indianapolis–Marion County storms, March 31, 2006 | 4 |
| Honeywell International, Inc., chlorine release, contaminated antimony pentachloride exposure, hydrogen fluoride release | 4 |
| Severe acute respiratory syndrome (SARS) complaint, Mid-Continent Airport | 4 |
| Teris LLC explosion and fire | 4 |
| Greensburg tornado | 4 |
| CTA Acoustics, Inc., combustible dust fire and explosions | 3 |
| CAI, Inc., and Arnel Company, Inc., confined vapor cloud explosion | 2 |
| Valero confined space entry | 2 |
| Third Coast Industries petroleum products facility incident | 2 |
| Holly tornado | 2 |
| Fairfax, Virginia, tanker fire | 1 |
| Morton International, Inc., chemical manufacturing incident | 1 |
| Chicago tanker fire | 1 |
| Herrig Brothers Feather Creek Farm propane tank explosion | 1 |
| West Pharmaceutical Services, Inc., dust explosion | 1 |
| First Chemical Corp. explosion and fire | 1 |
| Total | 1,213 |

APPENDIX F

# List of After-Action Reports Reviewed and Analyzed

*After Action Report SARS Complaint, Mid-Continent Airport June 12, 2003.*

*After Action Report,* Incident Name: Topanga Fire CA-LAC208724 Sept. 28–Oct. 6, 2005.

*After Action Report,* Incident Name: Holly Tornado, March 28, 2007.

*After Action Report,* Incident Name: City of Alamosa Salmonella outbreak, March–April 2008.

*After Action Report,* Incident Name: Weld County Tornados May 22, 2008, July 2008.

Aiken County Sheriff's Office, *Aiken County Sheriff's Office After-Action Report,* Incident Name: Graniteville Train Wreck January 2005.

Butler, Brett, *Memorandum Tanker Fire After Action Report,* Office of Emergency Management, Arlington, Va., February 2005.

California Task Force 3, *Oklahoma City US&R After-Action Report,* 2000.

Chatham Emergency Management Agency, *After Action Report Imperial Sugar Dixie Crystal Plant,* Savannah, Ga., February 7, 2008.

City of Charleston West Virginia, Office of Emergency Services and Homeland Security, *After Action Report,* Incident Name: DuPont Crisis Drill Train Derailment and Fire, June 14, 2007.

City of Seattle, *After-Action Report for February 28, 2001 Nisqually Earthquake,* Prepared by the Disaster Management Committee, July 2001.

Cook, John Lee, Jr., *Tire Recycling Facility Fire Nebraska City, Nebraska,* Technical Report Series, Report 145, Federal Emergency Management Agency, U.S. Fire Administration, 2002.

Copeland, Tom D., *Industrial Plastics Fire: Major Triage Operation Flint Township, Michigan (November 29, 1988),* Technical Report Series, Report 025, Federal Emergency Management Agency, U.S. Fire Administration, no date.

CSB—*See* U.S. Chemical Safety and Hazard Investigation Board.

Custer, Richard L. P., *Swimming Pool Chemical Plant Fire Springfield, Massachusetts (June 17, 1988),* Technical Report Series, Report 027, Federal Emergency Management Agency, U.S. Fire Administration, no date.

EG&G Technical Services, Inc., *2007 County of San Diego Firestorms After Action Report,* February 2007.

Environmental Protection Agency, *Norfolk Southern Derailment Graniteville, South Carolina,* presentation, no date.

EOC Planning Sub-Committee, *2002 Winter Olympics After Action Report,* April 4, 2002.

Gordon, Susan, "Review of Wash. Chlorine Incident Faults Firefighters," *FireRescue News*, March 11, 2008.

Governor's Office of Emergency Services, *San Simeon Earthquake After Action Report*, Prepared by OES Planning and Technological Assistance Branch, November 2004.

Graniteville-Vaucluse-Warrenville Fire Department, *Graniteville-Vaucluse-Warrenville Fire Department After Action Report*, Incident Name: Graniteville Train Wreck, January 2005.

Grays Harbor County Public Health and Social Services Department and Environmental Health Division, *After-Action Report Windstorm Response December, 2007—A Report to the Board of Health*, January 2008.

Hoff, Chris, *After Action Review Rollup Lessons Learned Center*, Incident Name: Battle Creek Complex, July 2007.

Indianapolis and Marion County Civil Defense, *The Indiana State Fairgrounds Coliseum Disaster October 31, 1963*, no date.

*Indianapolis–Marion County Emergency Management Storm Event March 31, 2006 After Action Report*, Prepared by Indianapolis–Marion County Emergency Management Division, May 2006, not available to the general public.

*Indianapolis–Marion County Emergency Management Storm Event April 2, 2006 After Action Report*, Prepared by Indianapolis–Marion County Emergency Management Division, May 2006, not available to the general public.

Jennings, Charles, *Gasoline Tanker Incidents in Chicago, Illinois and Fairfax County, Virginia (March 30, 1989 and May 29, 1989) Case Studies in Hazardous Materials Planning*, Technical Report Series, Report 032, Federal Emergency Management Agency, U.S. Fire Administration, no date.

Kailes, June Isaacson, *Southern California Wildfires After Action Report*, Center for Disability Issues and the Health Professions, September 2008.

Kaszniak, Mark, and John Vorderbrueggen, "Runaway Chemical Reaction Exposes Community to Highly Toxic Chemicals," *Journal of Hazardous Materials*, Vol. 159, 2008, pp. 2–12.

Mason, Steve, *Final Report*, Incident Name: Teris LLC Explosion and Fire, El Dorado, Arkansas, EPA Region 6, Emergency Readiness Team, Response and Prevention Branch, March 2005.

Mason, Steve, *Final Report*, Incident Name: Union Pacific/Burlington Northern Train Derailment, Macdona, TX, EPA Region 6, Emergency Readiness Team, Response and Prevention Branch, August 2004.

National Incident Management Organization, *Final Narrative "Lessons Learned" May 7 to June 12, 2007*, Incident Name: Greensburg Kansas Tornado FEMA Assist, FEMA 1699DR-KS KS-FEM-000197, Prepared by Boise Incident Management Team, 2007.

National Transportation Safety Board, *Railroad Accident Brief*, Accident Number: DCA-01-Mr-044, 2001.

National Transportation Safety Board, *Derailment of Norfolk Southern Railway Company Train 68QB119 with Release of Hazardous Materials and Fire New Brighton, PA October 20, 2006*, Accident Report NTSB/RAR-08/02 PB2008-916302, May 2008.

Nordin, John, "Alberton Canyon Chlorine Rail Car Derailment," *The First Responder*, April 2007.

NTSB—*See* National Transportation Safety Board.

Orange County Fire Authority, *After Action Report Santiago Fire October 21–November 9, 2007*, A Report to the Orange County Fire Authority Board of Directors, no date.

Quarantelli, E. L., David C. Hutchinson, and Brenda D. Phillips, *Evacuation Behavior: Case Study of the Taft, Louisiana Chemical Tank Explosion Incident*, Miscellaneous Report #34, University of Delaware Disaster Research Center, May 1983.

Routley, J. Gordon, *Massive Leak of Liquified Chlorine Gas Henderson, Nevada (May 6, 1991)*, Technical Report Series, Report 052, Federal Emergency Management Agency, U.S. Fire Administration, no date.

Sensenig, Daryl and Patrick Simpson, *Chemical Fire in Apex, North Carolina (October 5–7, 2006)*, Technical Report Series, Report 163, Federal Emergency Management Agency, U.S. Fire Administration, no date.

Shane, Daniel M., *Federal On-Scene Coordinator's Report*, Incident Name: Westley Tire Fire Stanislaus County, California, September 22, 1999, U.S. Environmental Protection Agency Region IX, August 2000.

Stambaugh, Hollis, *Evacuation of Nanticoke, Pennsylvania Due to Metal Processing Plant Fire (March 24, 1987)*, Technical Report Series, Report 005, Federal Emergency Management Agency, U.S. Fire Administration, no date.

Stern, Jeff, *Fire Department Response to Biological Threat at B'nai B'rith Headquarters Washington, DC*, Technical Report Series, Report 114, Federal Emergency Management Agency, U.S. Fire Administration, April 1997.

Styron, Hilary C., *CSX Tunnel Fire Baltimore, MD July 2001*, Technical Report Series, Report 140, Federal Emergency Management Agency, U.S. Fire Administration, no date.

U.S. Chemical Safety and Hazard Investigation Board, *Investigation Report Explosives Manufacturing Incident*, Incident Site: Sierra Chemical Company Mustang, NV Jan. 7, 1998, Report No. 98-001-I-NV, 1998.

U.S. Chemical Safety and Hazard Investigation Board, *Investigation Report Chemical Manufacturing Incident*, Incident Site: Morton International, Inc. Paterson, NJ April 8, 1998, Report No. 1998-06-I-NJ, 1998.

U.S. Chemical Safety and Hazard Investigation Board, *Investigation Report Propane Tank Explosion*, Incident Site: Herrig Brothers Feather Creek Farm Albert City, IA April 9, 1998, Report No. 98-007-I-IA, 1998.

U.S. Chemical Safety and Hazard Investigation Board, *Investigation Report Refinery Fire Incident*, Incident Site: Tosco Avon Refinery Martinez, CA Feb. 23, 1999, Report No. 99-014-I-CA, 2001.

U.S. Chemical Safety and Hazard Investigation Board, *Investigation Report Thermal Decomposition Incident*, Incident Site: BP Amoco Polymers, Inc., Augusta, GA March 13, 2001, Report No. 2001-03-I-GA, 2002.

U.S. Chemical Safety and Hazard Investigation Board, *Investigation Report Refinery Incident*, Incident Site: Motiva Enterprises LLC Delaware City Refinery Delaware City, DE July 17, 2001, Report No. 2001-05-I-DE, October 2002.

U.S. Chemical Safety and Hazard Investigation Board, *Investigation Report Hydrogen Sulfide Poisoning*, Incident Site: Georgia-Pacific Naheola Mill Pennington, AL Jan. 16, 2002, Report No. 2002-01-I-AL, 2003.

U.S. Chemical Safety and Hazard Investigation Board, *Investigation Report Petroleum Products Facility Incident*, Incident Site: Third Coast Industries Friendswood, TX May 1, 2002, Report No. 2002-03-I-TX, 2003.

U.S. Chemical Safety and Hazard Investigation Board, *Investigation Report Chlorine Release*, Incident Site: DPC Enterprise, L.P. Festus, MO Aug. 14, 2002, Report No. 2002-04-I-MO, 2003.

U.S. Chemical Safety and Hazard Investigation Board, *Investigation Report Explosion and Fire*, Incident Site: First Chemical Corporation Pascagoula, MS Oct. 13, 2002, Report No. 2003-01-I-MS, 2003.

U.S. Chemical Safety and Hazard Investigation Board, *Investigation Report Vapor Cloud Deflagration and Fire*, Incident Site: BLSR Operating, LTD. Rosharon, TX Jan. 13, 2003, Report No. 2003-06-I-TX, 2003.

U.S. Chemical Safety and Hazard Investigation Board, *Investigation Report Dust Explosion*, Incident Site: West Pharmaceutical Services, Inc. Kinston, NC Jan. 29, 2003, Report No. 2003-07-I-NC, 2004.

U.S. Chemical Safety and Hazard Investigation Board, *Investigation Report Vent Collection System Explosion*, Incident Site: Technic Inc. Cranston, RI Feb. 7, 2003, Report No. 2003-08-I-RI, 2004.

U.S. Chemical Safety and Hazard Investigation Board, *Case Study Explosion at Biochemical Facility: Liquid Nitric Oxide Release*, Incident Site: Isotech Miami Township, OH Sept. 21, 2003, Report No. 2003-15-C-OH, 2004.

U.S. Chemical Safety and Hazard Investigation Board, *Investigation Report Chlorine Release, Contaminated Antimony Pentachloride Exposure, Hydrogen Fluoride Release*, Incident Site: Honeywell International, Inc. Baton Rouge, LA, Report No. 2003-13-I-LA, 2005.

U.S. Chemical Safety and Hazard Investigation Board, *Investigation Report Combustible Dust Fire and Explosion*, Incident Site: CTA Acoustics, Inc. Corbin, KY Feb. 20, 2003, Report No. 2003-09-I-KY, 2005.

U.S. Chemical Safety and Hazard Investigation Board, *Investigation Report Aluminum Dust Explosion*, Incident Site: Hayes Lemmerz International-Huntington, Inc. Huntington, IN Oct. 29, 2003, Report No. 2004-01-I-IN, 2005.

U.S. Chemical Safety and Hazard Investigation Board, *Investigation Report Toxic Chemical Vapor Cloud Release*, Incident Site: MFG Chemical, Inc. Dalton, GA April 12, 2004, Report No. 2004-09-I-GA, 2006.

U.S. Chemical Safety and Hazard Investigation Board, *Investigation Report Sterigenics*, Incident Site: Sterigenics Ontario, CA Aug. 19, 2004, Report No. 2004-11-I-CA, 2006.

U.S. Chemical Safety and Hazard Investigation Board, *Case Study Fire at Formosa Plastics Corporation: Evaluating Process Hazards*, Incident Site: Formosa Plastics Corporation Point Comfort, TX Oct. 6, 2005, Report No. 2006-01-I-TX, 2006.

U.S. Chemical Safety and Hazard Investigation Board, *Case Study Confined Space Entry—Worker and Would-be Rescuer Asphyxiated*, Incident Site: Valero Energy Corporation Refinery Delaware City, DE Nov. 5, 2005, Report No. 2006-02-I-DE, 2006.

U.S. Chemical Safety and Hazard Investigation Board, *Investigation Report Chlorine Release*, Incident Site: DPC Enterprises, L.P. Glendale, AZ Nov. 17, 2003, Report No. 2004-02-I-AZ, 2007.

U.S. Chemical Safety and Hazard Investigation Board, *Investigation Report Vinyl Chloride Monomer Explosion*, Incident Site: Formosa Plastics Corp. Illiopolis, IL April 23, 2004, Report No. 2004-10-I-IL, 2007.

U.S. Chemical Safety and Hazard Investigation Board, *Investigation Report Methanol Tank Explosion and Fire*, Incident Site: Bethune Point Wastewater Treatment Plant city of Daytona Beach, FL Jan. 11, 2006, Report No. 2006-03-I-FL, 2007.

U.S. Chemical Safety and Hazard Investigation Board, *Investigation Report Confined Vapor Cloud Explosion*, Incident Site: CAI, Inc. and Arnel Company, Inc. Danvers, MA Nov. 22, 2006, Report No. 2007-03-I-MA, 2008.

U.S. Chemical Safety and Hazard Investigation Board, *Investigation Report Little General Store—Propane Explosion*, Incident Site: Little General Store, Inc. Ghent, WV Jan. 30, 2007, Report No. 2007-04-I-WV, 2008.

Volusia County Emergency Management, *After Action Review*, Incident Name: Volusia County Terrorism Rail Exercise, April 2006.

# Bibliography


Afshartous, David, Yongtao Guan, and Anuj Mehrotra, "US Coast Guard Air Station Location with Respect to Distress Calls: A Spatial Statistics and Optimization Based Methodology," *European Journal of Operational Research,* Vol. 196, 2009, pp. 1086–1096.

Altay, Nezih, and Walter G. Green III, "OR/MS Research in Disaster Operations Management," *European Journal of Operational Research*, Vol. 175, 2006, pp. 475–493.

Arboleda, Carlos A., Dulcy M. Abraham, and Robert Lubitz, "Simulation as a Tool to Assess the Vulnerability of the Operation of a Health Care Facility," *Journal of Performance of Constructed Facilities*, Vol. 21, No. 4, August 1, 2007, pp. 302–312.

Argonne National Laboratory, "Temporary Shelter-in-Place as Protection Against a Release of Airborne Hazardous Material: Report of a Literature Search," March 16, 2001.

Ashland Fire and Rescue, *Standards of Coverage 2009*, Ashland, Ore., 2009. As of June 3, 2010: http://www.ashland.or.us/Files/Standard%20Of%20Cover%20-final%204-16-09.pdf

Ball, Michael O., and Feng L. Lin, "A Reliability Model Applied to Emergency Service Vehicle Location," *Operations Research*, Vol. 41, No. 1, Special Issue on Stochastic and Dynamic Models in Transportation, January–February 1993, pp. 18–36.

Barrett, A., *Mathematical Modeling and Decision Analysis for Terrorism Defense: Assessing Chlorine Truck Attack Consequences and Countermeasure Cost Effectiveness*, Ph.D. thesis, Department of Engineering and Public Policy, Carnegie Mellon University, Pittsburg, Pa., May 2009.

Beraldi, P., M. E. Bruni, and D. Conforti, "Designing Robust Emergency Medical Service Via Stochastic Programming," *European Journal of Operational Research*, Vol. 158, 2004, pp. 183–193.

Bigley, Gregory A., and Karlene H. Roberts, "The Incident Command System: High-Reliability Organizing for Complex and Volatile Task Environments," *Academy of Management Journal*, Vol. 44, No. 6, December 2001, pp. 1281–1299.

Boisvert, A., *Understanding Hazardous Materials, Operations, and Emergency Response*, Authorhouse: Bloomington, Ind., 2007.

Brittle, Stephen, "Emergency Response Issues: What Went Wrong in Graniteville," no date. As of June 4, 2010:
http://www.chemicalspill.org/railcar.html

Byers, M., M. Russell, and D. J. Lockey, "Clinical Care in the 'Hot Zone,'" *Emergency Medical Journal*, Vol. 25, 2008, pp. 108–112.

Chelst, Kenneth, and James P. Jarvis, "Estimating the Probability Distribution of Travel Times for Urban Emergency Service Systems," *Operations Research*, Vol. 27, No. 1, January–February 1979, pp. 199–204.

Chen, X., and F. B. Zhan, "Agent-Based Modeling and Simulation of Urban Evacuation: Relative Effectiveness of Simultaneous and Staged Evacuation Strategies," *Journal of the Operational Research Society*, Vol. 59, 2008, pp. 25–33.

Chilcott, R. P., "Chlorine: Incident Management," Health Protection Agency (UK), 2007.

Chlorine Institute Inc., *Chlorine: Effects on Health and the Environment*, 3rd ed., November 1999. As of May 27, 2010:
http://www.chlorineinstitute.org/files/PDFs/ChlorineEffectsOnHealth.pdf

CNN.com, "Iraq Gas Attack Makes Hundreds Ill," March 17, 2007. As of May 27, 2010:
http://www.cnn.com/2007/WORLD/meast/03/17/iraq.main/index.html

Conrad, S. H., T. Brown, and W. Beyeler, "A Dynamic Simulation Model of the Effects of Interdependent Infrastructures on Emergency Service Response," 20th International Conference of the System Dynamics Society (Albany, New York), Palermo, Italy, 2002.

CSB—*See* U.S. Chemical Safety and Hazard Investigation Board.

Dausey, David J., Anita Chandra, Agnes G. Schaefer, Ben Bahney, Amelia Haviland, Sarah Zakowski, and Nicole Lurie, "Measuring the Performance of Telephone-Based Disease Surveillance Systems in Local Health Departments," *American Journal of Public Health*, September 2008, Vol. 98, No. 9, pp. 1706–1711.

DeAtley, C., S. Allen, W. Hauda, P. DeHaven, and A. Stangby, *Jane's Mass Casualty Handbook: Pre-hospital Emergency Preparedness and Response*, 1st ed., Jane's Information Group, Surrey, UK, 2003.

DHS—*See* U.S. Department of Homeland Security.

DoD—*See* U.S. Department of Defense.

Donahue, Amy, and Robert V. Tuohy, "Lessons We Don't Learn: A Study of the Lessons of Disasters, Why We Repeat Them, and How We Can Learn Them," *Homeland Security Affairs*, Vol. 2, No. 2, July 2006. As of May 27, 2010:
http://www.hsaj.org/?home=2.2

Ebeling, Charles E., *An Introduction to Reliability and Maintainability Engineering*, New York, N.Y.: McGraw Hill, 1997.

EMAP—*See* Emergency Management Accreditation Program.

Emergency Management Accreditation Program, "EMAP Standard," September 2007.

Enthoven, Alain C., and K. Wayne Smith, *How Much Is Enough? Shaping the Defense Program, 1961–1969*, Santa Monica, Calif.: RAND Corporation, CB-403, 2005. As of May 26, 2010:
http://www.rand.org/pubs/commercial_books/CB403/

FAA—*See* Federal Aviation Administration.

Federal Aviation Administration, *FAA System Safety Handbook, Chapter 9: Analysis Techniques*, December 30, 2000.

Federal Emergency Management Agency, "2009 Federal Disaster Declarations," 2009a. As of May 27, 2010:
http://www.fema.gov/news/disasters.fema?year=2009

———, *The Federal Preparedness Report*, January 13, 2009b.

FEMA—*See* Federal Emergency Management Agency.

Fry, Michael J., Michael J. Magazine, and Uday S. Rao, "Firefighter Staffing Including Temporary Absences and Wastage," *Operations Research*, Vol. 54, No. 2, March–April 2006, pp. 353–365.

Georgiadou, Paraskevi S., Ioannis A. Papazoglou, Chris T. Kiranoudis, and Nikolaos C. Markatos, "Modeling Emergency Evacuation for Major Hazard Industrial Sites," *Reliability Engineering and System Safety*, Vol. 92, 2007, pp. 1388–1402.

Gordon, Susan, "Review of Wash. Chlorine Incident Faults Firefighters," *Fire Rescue 1 News*, March 3, 2008.

Han, Lee D., Fang Yuan, and Thomas Urbanik II, "What Is an Effective Evacuation Operation?" *Journal of Urban Planning and Development*, Vol. 133, No. 1, March 2007, pp. 3–8.

Hecht, Herbert, *Systems Reliability and Failure Prevention*, Norwood, Mass.: Artech House, Inc., 2003.

Horton, D. Kevin, Zahava Berkowitz, Gilbert S. Haugh, Maureen Orr, and Wendy Kaye, "Acute Public Health Consequences Associated with Hazardous Substances Released During Transit, 1993–2000," *Journal of Hazardous Materials*, B98, pp. 161–175, 2003.

Houghton, Brian, *Gearing Up and Getting There: Improving Local Response to Chemical Terrorism*, Santa Monica, Calif.: RAND Corporation, RGSD-181, 2004. As of May 27, 2010: http://www.rand.org/pubs/rgs_dissertations/RGSD181/

Hupert, Nathaniel, Alvin I. Mushlin, and Mark A. Callahan, "Modeling the Public Health Response to Bioterrorism: Using Discrete Event Simulation to Design Antibiotic Distribution Centers," *Medical Decision Making*, Vol. 22, 2002, pp. S17–S25.

Iannoni, Ana Paula, Reinaldo Morabito and Cem Saydam, "A Hypercube Queueing Model Embedded into a Genetic Algorithm for Ambulance Deployment on Highways," *Annals Operations Research*, Vol. 157, 2008, pp. 207–224.

Institute of Medicine, National Academies of Sciences and Engineering, *Chemical and Biological Terrorism: Research and Development to Improve Civilian Medical Response*, National Academies Press: Washington, D.C., 1999.

International Association of Fire Chiefs, *Terrorism Response: A Checklist and Guide for Fire Chiefs*, no date.

Jackson, Brian A., *The Problem of Measuring Emergency Preparedness: The Need for Assessing "Response Reliability" as Part of Homeland Security Planning*, Santa Monica, Calif.: RAND Corporation, OP-234-RC, 2008. As of May 27, 2010: http://www.rand.org/pubs/occasional_papers/OP234/

Jackson, Brian A., John C. Baker, M. Susan Ridgely, James T. Bartis, and Herbert I. Linn, *Protecting Emergency Responders, Volume 3: Safety Management in Disaster and Terrorism Response*, Santa Monica, Calif.: RAND Corporation, MG-170-NIOSH, 2004. As of May 27, 2010: http://www.rand.org/pubs/monographs/MG170/

Jackson, Brian A., and David R. Frelinger, *Emerging Threats and Security Planning: How Should We Decide What Hypothetical Threats to Worry About?* Santa Monica, Calif.: RAND Corporation, OP-256-RC, 2009a. As of May 27, 2010: http://www.rand.org/pubs/occasional_papers/OP256/

———, *Understanding Why Terrorist Operations Succeed or Fail?* Santa Monica, Calif.: RAND Corporation, OP-257-RC, 2009b. As of May 27, 2010: http://www.rand.org/pubs/occasional_papers/OP257/

Jackson, Brian A., D. J. Peterson, James T. Bartis, Tom LaTourrette, Irene T. Brahmakulam, Ari Houser, and Jerry M. Sollinger, *Protecting Emergency Responders: Lessons Learned from Terrorist Attacks*, Santa Monica, Calif.: RAND Corporation, CF-176-OSTP, 2002. As of May 27, 2010: http://www.rand.org/pubs/conf_proceedings/CF176/

Joseph, Giby, "Case Study Chlorine Transfer Hose Failure," *Journal of Hazardous Materials*, Vol. 115, 2004, pp. 119–125.

Jotshi, Arun, Qiang Gong, and Rajan Batta, "Dispatching and Routing of Emergency Vehicles in Disaster Mitigation Using Data Fusion," *Socio-Economic Planning Sciences*, Vol. 43, 2009, pp. 1–24.

Kanno, Taro, and Kazuo Furuta, "Resilience of Emergency Response Systems," no date. As of May 27, 2010:
http://www.resilience-engineering.org/REPapers/Kanno_Furuta_R.pdf

Kaplan, Edward H., and Johan Walden, "Situational Awareness in a Bioterror Attack Via Probabilistic Modeling," in M.S. Green et al. (eds.), *Risk Assessment and Risk Communication Strategies in Bioterrorism Preparedness*, Springer, 2007, pp. 31–44.

Kelley, Charles T., Jr., *The Impact of Equipment Availability and Reliability on Mission Outcomes: An Initial Look*, Santa Monica, Calif.: RAND Corporation, DB-423-A, 2004. As of May 27, 2010:
http://www.rand.org/pubs/documented_briefings/DB423/

Kolesar, Peter J., Kenneth L. Rider, Thomas B. Crabill, and Warren E. Walker, "A Queuing-Linear Programming Approach to Scheduling Police Patrol Cars," *Operations Research*, Vol. 23, No. 6, November–December 1975, pp. 1045–1062.

Larson, Richard C., Michael D. Metzger, and Michael F. Cahn, *Emergency Response for Homeland Security: Lessons Learned and the Need for Analysis*, Los Angeles, Calif.: Center for Risk and Economic Analysis of Terrorism Events, University of Southern California, 2004. As of May 27, 2010:
http://www.usc.edu/dept/create/assets/001/50756.pdf

Lee, Eva K., Siddhartha Maheshwary, Jacquelyn Mason, and William Glisson, "Decision Support System for Mass Dispensing of Medications for Infectious Disease Outbreaks and Bioterrorist Attacks," *Annals of Operations Research*, Vol. 148, 2006, pp. 25–53.

Miskel, James F., *Disaster Response and Homeland Security: What Works, What Doesn't*, Stanford, Calif.: Stanford University Press, 2008.

Modarres, Mohammad, Mark Kaminskiy, and Vasiliy Krivtsov, *Reliability Engineering and Risk Analysis: A Practical Guide*, New York, N.Y.: Marcel Dekker, 1999.

Morgan, Millett Granger, and Max Henrion, *Uncertainty: A Guide to Dealing with Uncertainty in Quantitative Risk and Policy Analysis*, New York, N.Y.: Cambridge University Press, 1990.

National Fire Protection Association, *NFPA 471—Recommended Practice for Responding to Hazardous Materials Incidents*, 1997.

———, *NFPA 1710—Standard for the Organization and Deployment of Fire Suppression Operations, Emergency Medical Operations, and Special Operations to the Public by Career Fire Departments*, 2001.

———, *NFPA 1600—Standard on Disaster/Emergency Management and Business Continuity Programs*, 2007.

National Research Council Subcommittee on Acute Exposure Guideline Levels, *Acute Exposure Guideline Levels for Selected Airborne Chemicals, Vol. 4*, Washington, D.C.: National Academies Press, 2004.

National Transportation Safety Board, *Collision of Norfolk Southern Freight Train 192 with Standing Norfolk Southern Local Train P22 with Subsequent Hazardous Materials Release at Graniteville, South Carolina January 6, 2005*, Railroad Accident Report, NTSB/RAR-05/04, Washington, D.C., Nov. 29, 2005.

Nelson, Christopher, Edward W. Chan, Anita Chandra, Paul Sorensen, Henry H. Willis, Katherine Comanor, Hayoung Park, Karen A. Ricci, Leah B. Caldarone, Molly Shea, John A. Zambrano, and Lydia Hansell, *Recommended Infrastructure Standards for Mass Antibiotic Dispensing*, Santa Monica, Calif.: RAND Corporation, TR-553-DHHS, 2008. As of May 27, 2010:
http://www.rand.org/pubs/technical_reports/TR553

Nelson, Christopher, Nicole Lurie, and Jeffery Wasserman, "Assessing Public Health Emergency Preparedness: Concepts, Tools, and Challenges," *Annual Review of Public Health*, Vol. 28, 2007a, pp. 12.1–12.18.

Nelson, Christopher, Nicole Lurie, Jeffery Wasserman, and Sara Zakowski, "Conceptualizing and Defining Public Health Emergency Preparedness," *American Journal of Public Health*, Vol. 97, No. S1, 2007b, pp. S9–S11.

NFPA—*See* National Fire Protection Association.

Norfolk Southern, "Norfolk Southern Reaches Agreement with Avondale Mills to Settle Claims From Graniteville Accident," press release, April 7, 2008. As of May 27, 2010:
http://www.nscorp.com/nscportal/nscorp/Media/News%20Releases/2008/agreement.html

NTSB—*See* National Transportation Safety Board.

Office of Risk Management and Analysis (RMA), "Attack Path for Standoff Attacks Against Chlorine Facilities, Risk Assessment Process for Investment Decisionmaking (RAPID)," August 2008.

O'Reilly, Gerard, Huseyin Uzunalioglu, Stephen Conrad, and Walt Beyeler, "Inter-Infrastructure Simulations Across Telecom, Power, and Emergency Services," paper presented at the Design of Reliable Communication Networks—5th International Workshop, 2005.

Pal, Raktim, and Indranil Bose, "An Optimization Based Approach for Deployment of Roadway Incident Response Vehicles with Reliability Constraints," *European Journal of Operational Research*, Vol. 198, 2009, pp. 452–463.

Peeta, Srinivas, F. Sibel Salman, Dilek Gunnec, and Kannan Viswanath, "Pre-Disaster Investment Decisions for Strengthening a Highway Network," *Computers and Operations Research*, Vol. 37, 2010, pp. 1708–1719.

Perrow, Charles, *Normal Accidents: Living with High-Risk Technologies*, Princeton, N.J.: Princeton University Press, 1999

Phoenix, AZ, Fire Department, (n.d.), *Standard Operating Procedures, Volume 2*, no date. As of May 27, 2010:
http://phoenix.gov/fire/forfiredepts.html

Public Law 109-295, The Post-Katrina Emergency Management Reform Act, October 4, 2006.

Raber, Ellen, Joy M. Hirabayashi, Saverio P. Mancieri, Alfred L. Jin, Karen J. Folks, Tina M. Carlsen, and Pete Estacio, "Chemical and Biological Agent Incident Response and Decision Process for Civilian and Public Sector Facilities," *Risk Analysis*, Vol. 22, No. 2, 2002, pp. 195–202.

Rawls, Carmen G., and Mark A. Turnquist, "Pre-Positioning of Emergency Supplies for Disaster Response," Transportation Research: Part B, Vol. 44, 2010, pp. 521–534.

Revelle, Charles, and Kathleen Hogan, "The Maximum Reliability Location Problem and α-Reliable p-Center Problem: Derivatives of the Probabilistic Location Set Covering Problem," *Annals of Operations Research*, Vol. 18, 1989, pp. 155–174.

Richards, Gregory, "Norfolk Southern Set to Fight EPA Lawsuit, CEO Tells Shareholders," *The Virginian-Pilot*, May 9, 2008. As of May 27, 2010:
http://hamptonroads.com/2008/05/norfolk-southern-set-fight-epa-lawsuit-ceo-tells-shareholders

Rudolph, Jenny W., and Nelson P. Repenning, "Disaster Dynamics: Understanding the Role of Quantity in Organizational Collapse," *Administrative Science Quarterly*, Vol. 47, 2002, pp. 1–30.

Sagan, Scott D., *The Limits of Safety: Organizations, Accidents, and Nuclear Weapons*, Princeton, N.J.: Princeton University Press, 1993.

Sanders, Jane, "Chlorine's Casualties and Counsel," Research Horizons, Georgia Institute of Technology, Fall 2006.

Silva, F., and D. Serra, "Locating Emergency Services with Different Priorities: The Priority Queuing Covering Location Problem," *Journal of the Operational Research Society*, Vol. 59, 2008, pp. 1229–1238.

Simpson, N. C., and P. G. Hancock, "Fifty Years of Operational Research and Emergency Response," *Journal of the Operational Research Society*, Vol. 60, 2009, pp. S126–S139.

Sorensen, Paul, and Richard Church, "Integrating Expected Coverage and Local Reliability for Emergency Medical Services Location Problems," *Socio-Economic Planning Sciences*, Vol. 44, 2010, pp. 8–18.

Stepanov, Alexander, and James MacGregor Smith, "Multi-Objective Evacuation Routing in Transportation Networks," *European Journal of Operational Research*, Vol. 198, 2009, pp. 435–446.

U.S. Army, "Chemical Accident or Incident Response and Assistance (CAIRA) Operations," Pamphlet 50–6, March 26, 2003.

———, *Failure Modes, Effects and Criticality Analysis (FMECA) for Command, Control, Communications, Computer, Intelligence, Surveillance, and Reconnaissance (C4ISR) Facilities*, TM 5-698-4, September 29, 2006.

U.S. Chemical Safety and Hazard Investigation Board, "Completed Investiagtions," web page, no date, accessed April 22, 2009. As of June 2, 2010:
http://www.csb.gov/investigations/investigations.aspx?Type=2&F_%20All=y

———, "Investigation Report Chlorine Release DPC Enterprises, L.P. Glendale, Arizona November 17, 2003," Report No. 2004-02-I-AZ, Feb. 2007.

U.S. Department of Defense, *Procedures for Performing a Failure Mode, Effects and Criticality Analysis*, MIL-STD-1629A, November 24, 1980.

U.S. Department of Homeland Security, *National Planning Scenarios: Created for Use in National, Federal, State, and Local Homeland Security Preparedness Activities*, Version 20.1, April 2005. As of May 27, 2010:
http://media.washingtonpost.com/wp-srv/nation/nationalsecurity/earlywarning/NationalPlanningScenariosApril2005.pdf

———, *National Planning Scenario 5: Chemical Attack—Blister Agent*, Version 21.3, March 2006a.

———, *National Planning Scenario 6: Chemical Attack—Toxic Industrial Chemicals*, Version 21.3, March 2006b.

———, *National Planning Scenario 7: Chemical Attack—Nerve Agent*, Version 21.3, March 2006c.

———, *National Planning Scenario 8: Chemical Attack—Chlorine Tank Explosion*, Version 21.3, March 2006d.

———, "Chemical Weapons Attack Tree for DHS National Planning Scenario 7," Version 3, August 2006e.

———, *Universal Task List*, February 2007a.

———, *Target Capabilities List*, September 2007b.

———, "Daily Open Source Infrastructure Report for 22 January 2008," 2008a. As of May 27, 2010:
http://osd.gov.com/osd/200801_January/DHS_Daily_Report_2008-01-22.pdf

———, *National Incident Management System*, December 2008b.

U.S. Nuclear Regulatory Commission, *Fault Tree Handbook*, NUREG-0492, January 1981.

Wein, Lawrence M., David L. Craft, and Edward H. Kaplan, "Emergency Response to a Smallpox Attack: The Case for Mass Vaccination," *Proceedings of the National Academy of Sciences*, Vol. 99, No. 16, 2002, pp. 10935–10940.

———, "Emergency Response to an Anthrax Attack," *Proceedings of the National Academy of Sciences*, Vol. 100, No. 7, 2003, pp. 4346–4351.

Willis, Henry H., Christopher Nelson, Shoshana R. Shelton, Andrew M. Parker, John A. Zambrano, Edward W. Chan, Jeffrey Wasserman, and Brian A. Jackson, *Initial Evaluation of the Cities Readiness Initiative*, Santa Monica, Calif.: RAND Corporation, TR-640-CDC, 2009. As of May 27, 2010:
http://www.rand.org/pubs/technical_reports/TR640/

Withers, R. M. J., and F. P. Lees, "The Assessment of Major Hazards: The Lethal Toxicity of Chlorine, Part 2: Model of Toxicity to Man," *Journal of Hazardous Materials*, Vol. 12, No. 3, December 1985, pp. 283–302.

World Health Organization, *Manual for the Public Health Management of Chemical Incidents*, Geneva, Switzerland: WHO Press, 2009.

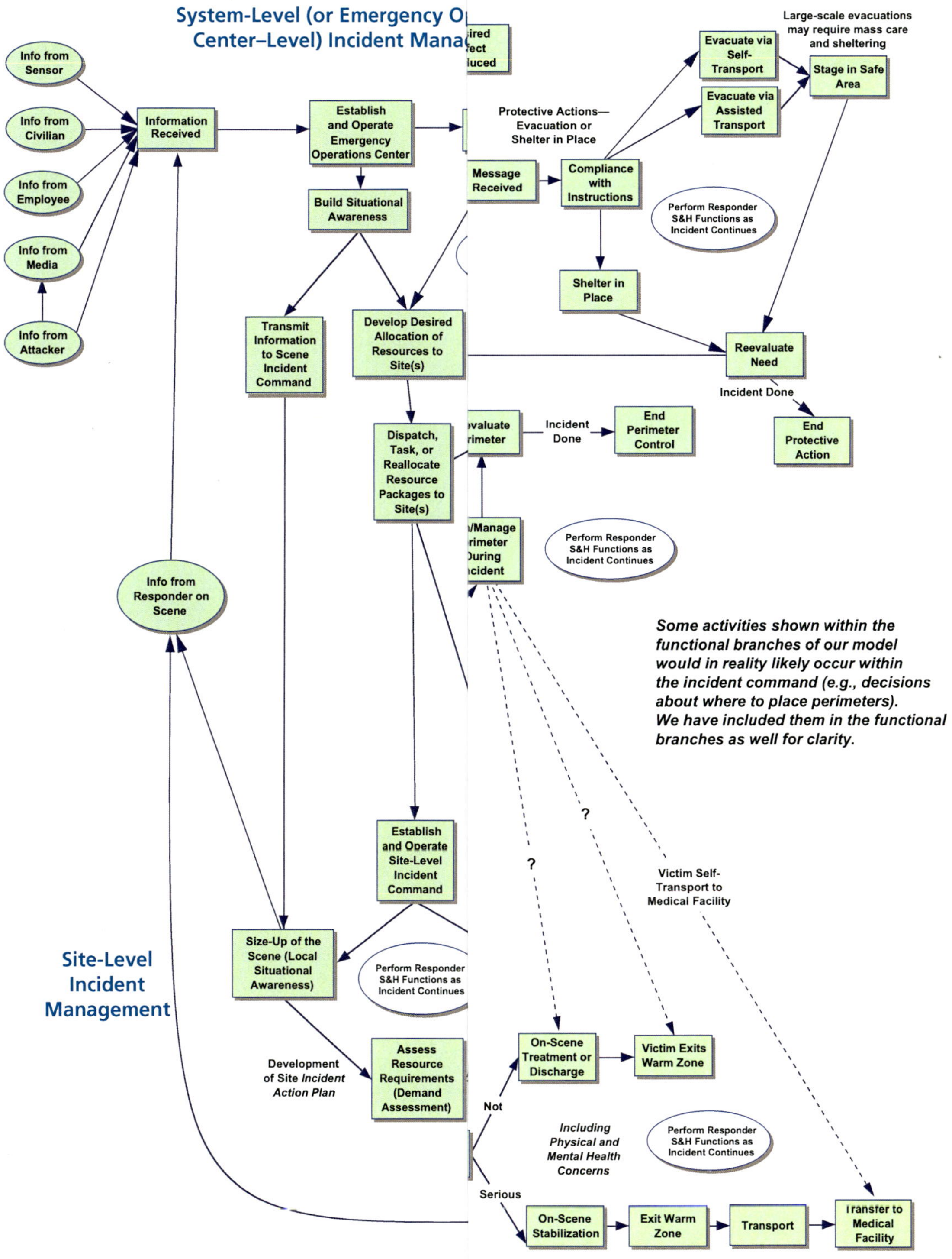

System-Level (or Emergency O
Center–Level) Incident Mana
Info from Sensor
Info from Civilian
Info from Employee
Info from Media
Info from Attacker
Information Received
Establish and Operate Emergency Operations Center
Build Situational Awareness
Transmit Information to Scene Incident Command
Develop Desired Allocation of Resources to Site(s)
Dispatch, Task, or Reallocate Resource Packages to Site(s)
Info from Responder on Scene
Establish and Operate Site-Level Incident Command
Size-Up of the Scene (Local Situational Awareness)
Perform Responder S&H Functions as Incident Continues
Development of Site Incident Action Plan
Assess Resource Requirements (Demand Assessment)
Site-Level Incident Management
ired
fect
luced
Protective Actions— Evacuation or Shelter in Place
Message Received
Compliance with Instructions
Evacuate via Self-Transport
Evacuate via Assisted Transport
Large-scale evacuations may require mass care and sheltering
Stage in Safe Area
Perform Responder S&H Functions as Incident Continues
Shelter in Place
Reevaluate Need
Incident Done
End Protective Action
valuate
rimeter
Incident Done
End Perimeter Control
/Manage
rimeter
During
ncident
Perform Responder S&H Functions as Incident Continues
Some activities shown within the functional branches of our model would in reality likely occur within the incident command (e.g., decisions about where to place perimeters). We have included them in the functional branches as well for clarity.
?
?
Victim Self-Transport to Medical Facility
On-Scene Treatment or Discharge
Victim Exits Warm Zone
Not
Including Physical and Mental Health Concerns
Perform Responder S&H Functions as Incident Continues
Serious
On-Scene Stabilization
Exit Warm Zone
Transport
Transfer to Medical Facility